Mechanical Engineering Series

Frederick F. Ling
Series Editor

George A. Costello

Theory of Wire Rope

With 46 Figures

Springer-Verlag
New York Berlin Heidelberg
London Paris Tokyo Hong Kong

George A. Costello
Department of Theoretical and Applied Mechanics, University of Illinois at Urbana-Champaign, Urbana, IL 61801, USA

Series Editor

Frederick F. Ling
Director, Columbia Engineering Productivity Center, and Professor, Department of Mechanical Engineering, Columbia University, New York, NY 10027-6699; and Distinguished William Howard Hart Professor Emeritus, Department of Mechanical Engineering, Aeronautical Engineering and Mechanics, Rensselaer Polytechnic Institute, Troy, NY 12180-3590, USA

Library of Congress Cataloging-in-Publication Data
Costello, George A. (George Albert)
Theory of wire rope/by George A. Costello.
p. cm.—(Mechanical engineering series)
Includes bibliographical references.
ISBN 0-387-97189-0 (alk. paper)
1. Wire-rope. 2. Wire rope—Testing. I. Title. II. Series: Mechanical engineering series (Berlin, Germany)
TA492.W8C67 1990
671.8′4—dc20 89-26289

Printed on acid-free paper.

Typeset by Asco Trade Typesetting Ltd., Hong Kong.
Printed and bound by Edwards Brothers, Inc., Ann Arbor, Michigan.
Printed in the United States of America.

9 8 7 6 5 4 3 2 1

ISBN 0-387-97189-0 Springer-Verlag New York Berlin Heidelberg
ISBN 3-540-97189-0 Springer-Verlag Berlin Heidelberg New York

To my wife, Jean, and our three daughters,
Suzanne, Elizabeth, and Lisa

Series Preface

Mechanical engineering, an engineering discipline borne of the needs of the industrial revolution, is once again asked to do its substantial share in the call for industrial renewal. The general call is urgent as we face profound issues of productivity and competitiveness that require engineering solutions, among others. The Mechanical Engineering Series is a new series, featuring graduate texts and research monographs, intended to address the need for information in contemporary areas of mechanical engineering.

The series is conceived as a comprehensive one that will cover a broad range of concentrations important to mechanical engineering graduate education and research. We are fortunate to have a distinguished roster of consulting editors, each an expert in one of the areas of concentration. The names of the consulting editors are listed on the first page of the volume. The areas of concentration are applied mechanics, biomechanics, computational mechanics, dynamic systems and control, energetics, mechanics of materials, processing, thermal science, and tribology.

Professor Leckie, the consulting editor for applied mechanics, and I are pleased to present the third volume of the series: *Theory of Wire Rope* by Professor Costello. The selection of this volume underscores again the interest of the Mechanical Engineering Series to provide our readers with topical monographs as well as graduate texts.

New York, New York *Frederick F. Ling*

Preface

This book, as the title indicates, is concerned with the various theories of wire rope. During recent years, considerable progress has been made in the development of models used to predict the response of wire rope. Since there are so many parameters that can vary in the construction of rope, such models can be used to determine the effects of possible variations of the parameters on the performance of a rope.

A list of the uses of wire rope is almost endless. Recent research into the possible use of wire strands as braces for teeth is one such example. Wire rope is used to lower men underground as deep as 16,000 ft in the gold mines of South Africa. This is, of course, accomplished by more than one lift, since the weight of the rope would be excessive in a single lift. One such rope used in a shaft that runs over several sheaves is 9.3 mi long and weighs 110 tn. Many power lines can be regarded as a strand consisting of aluminum wires twisted around a steel center wire. Wire strands are used as cords to strengthen rubber tires. Wire rope is also being considered in superconductivity applications.

The basic components and construction of wire rope are treated in Chapter 1. Although there are many different types of construction, a rope is generally regarded as having three components: (1) wires that form the strand, (2) a core, and (3) multiwire strands that are helically wrapped around the core.

Chapter 2 begins with an investigation of the kinematics of a thin wire. The equations of equilibrium are then derived for a wire, and the relations between the internal loads and deformation are presented. The wires are then placed together to form a strand, in Chapter 3, where consideration is given to the static response of a strand subjected to an axial tensile force and an axial twisting moment. The bending of a strand is next investigated, and the results are applied to a strand passing over a sheave. Expressions are presented for the axial wire stresses in the above cases.

Once the static response of a strand is determined, the results are extended to wire rope, in Chapter 4. An independent wire rope core (IWRC) is considered first, and then more complex cross sections are investigated. Expressions are again presented for the stresses in the rope, and plots depicting the maximum axial wire stresses in the individual wires are drawn.

Chapter 5 presents some aspects of friction in rope. The effective length of a fractured wire in a rope is discussed. This effective length is based on the contact loads between the wires, Coulomb-type friction, and an invocation of Saints–Venant's principle. Friction is also considered in the bending of a simple strand under tension.

In Chapter 6 some aspects of wire rope testing are considered. Strength test results are greatly enhanced by the use of dimensional analysis when the size effect is taken into account. Fatigue behavior in bending, when the size effect is accounted for, is also discussed.

The interesting phenomenon of birdcaging in wire rope is discussed in Chapter 7. A *bird cage* is a term often used to describe the permanent appearance of a wire rope forced into compression. Such damage, of course, renders the rope useless.

Chapter 8 considers the effects of rotation on the load-carrying capacity of a wire rope. If a rope is allowed to rotate, the failure load can be considerably reduced, especially when the ends of the ropes are spliced.

Most of the work presented in this book is based on research that my colleagues and I have performed for the last 16 years at the University of Illinois, Urbana-Champaign. I would like to thank especially Professor J.W. Phillips for his many contributions to the work in the form of ideas, computer plots, drawings, and photographs. The graduate students involved in the research were Dr. S.K. Sinha, Dr. G.J. Butson, Dr. S.A. Velinsky, Dr. C.H. Chien, Dr. R.A. LeClair, Mr. T.A. Conway, and Mr. C.C. Lin. A special note of thanks should go to Mr. E.H. Skinner and Mr. G.L. Anderson of the Spokane Research Center, Bureau of Mines, for their support in much of this work. I would also like to thank Ms. Jan Weaver for her outstanding typing.

Urbana, Illinois *George A. Costello*

Contents

1
Introduction

1.1 Basic Components

A property common to structural elements such as rope, yarn, cord, cable, and strand is their ability to resist relatively large axial loads in comparison to bending and torsional loads. Rope [1, 2],* because of this property, is one of the oldest tools that humans have used in their efforts to produce a better life for themselves. A copper cable found in the ruins of Nemeveh near Babylon indicates that wire rope was used as a structural element in about 700 B.C. Sayenga has written an excellent history of the American wire rope industry [3].

The basic element of a wire rope is, as the name implies, a single thin metallic wire. The various components of a wire rope are shown in Figure 1.1, where the rope is constructed by laying several strands around a core [1]. The core may be either wire rope, natural fibers, or polypropylene. The strands themselves have a center wire that is the axial member around which the individual metallic wires are wrapped helically. It should be mentioned that the major portion of the load acting on a rope is carried by the strands. The main purpose of the core is to provide proper support for the strands under normal bending and loading conditions. Most of the ropes used in the United States are designed with six strands, and there strands are comprised of any number of wires.

1.2 Identification and Construction

In addition to its components, wire rope is identified by its construction in the way its wires have been laid to form the strands and in the way its strands have been laid around the core. Figure 1.2 illustrates (a) a right regular lay, (b) a left regular lay, (c) a right lang lay, (d) a left lang lay, and (e) right alternate lay wire ropes. The wires in the regular lay appear to line up with the axis of

* Numbers in brackets denote references listed at the end of the book.

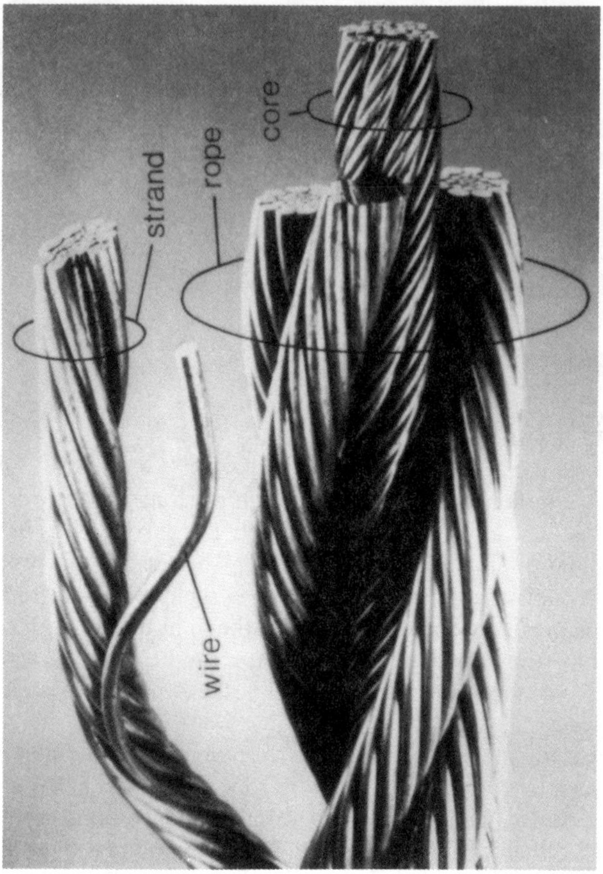

FIGURE 1.1. Wire rope.

the rope, whereas in the lang lay rope, the wires form an angle with the axis of the rope.

Figure 1.3 shows several basic cross-sectional constructions, around which stranded wire ropes are made. Wire rope is generally identified by a reference to its number of strands, as well as to the number and geometric arrangement of the wires in the strand. In addition, most of the rope produced today is performed, that is, the strands are permanently shaped, before fabrication into a rope, into the helical form they will assume in the rope. Figures 1.2 and 1.3 are reproduced from the *Wire Rope Users Manual* [4], an excellent publication produced by the American Iron and Steel Institute and the Wire Rope Technical Board. For a more complete description of wire rope identification and construction, the interested reader is referred to the above-mentioned work.

FIGURE 1.2. Typical wire rope lays: (a) right regular lay, (b) left regular lay, (c) right lang lay, (d) left lang lay, (e) right alternate lay.

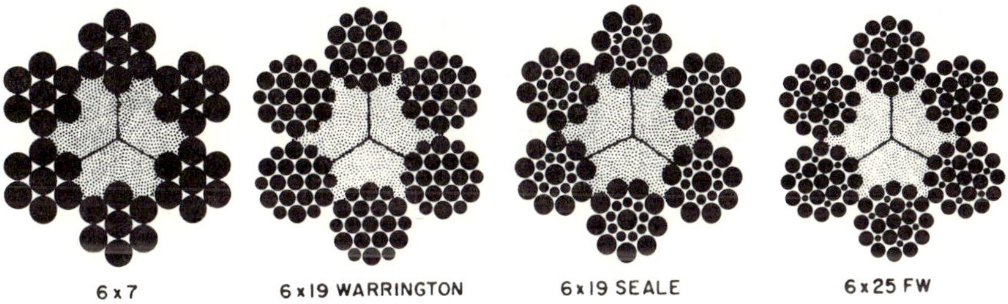

FIGURE 1.3. Basic cross-sectional constructions.

2
Equilibrium of a Thin Wire

2.1 Kinematics of a Thin Wire

Consider a curved thin wire that is initially unstressed and that has a uniform cross section. A thin wire can be defined as a wire in which the maximum characteristic length of the cross section, that is, its diameter or diagonal, etc., is small compared to the length of the wire and the radius of curvature of the centerline of the wire.

Let the cross section have two axes of symmetry as shown in Figure 2.1. The axes of symmetry will be the principal axes of the cross section. Consider now a system of axes A, B, and C, where the C-axis is tangent to the centroidal axis at point P and the A- and B-axes are perpendicular to the C-axis and are in the direction of the principal axes of the cross section. Let three infinitesimal line elements of the wire a, b, and c issue from P in the directions of A, B, and C.

Let the wires now be deformed. The three line elements a, b, and c of the wire, in general, do not remain perpendicular to each other. The elements a and c in the deformed state do, however, form a plane. Let the z-axis be in the direction of the deformed element c and let the x-axis be perpendicular to the z-axis and lie in the plane of the deformed elements a and c. Let the y-axis be perpendicular to both x and z so that the x-, y-, and z-axes form a right-handed orthogonal system. Hence, at each point along the deformed centroidal axis of the thin wire, a set of x-, y-, and z-axes exists. The system of axes constructed as described above for any point on the deformed axis is called the *principal torsion-flexure axes* [5].

Let the origin P' of the frame of the x-, y-, and z-axes move with a unit velocity along the deformed centroidal axis. This rotating frame will possess an angular velocity vector, $\vec{\omega}$. The projections or components of this vector in the x, y, and z directions will be defined as the components of the curvature κ, κ' and the twist per unit length τ. In the undeformed wire, the components of the angular velocity vector $\vec{\omega}_0$ (again moving with a unit velocity) projected on the A-, B-, and C-axes will define the undeformed components of curvature κ_0, κ'_0 and the twist per unit length τ_0.

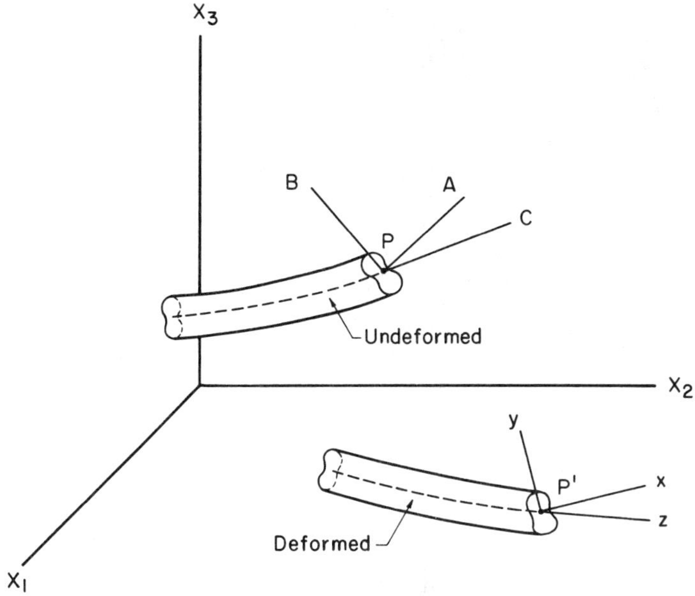

FIGURE 2.1. Undeformed and deformed curved thin wire.

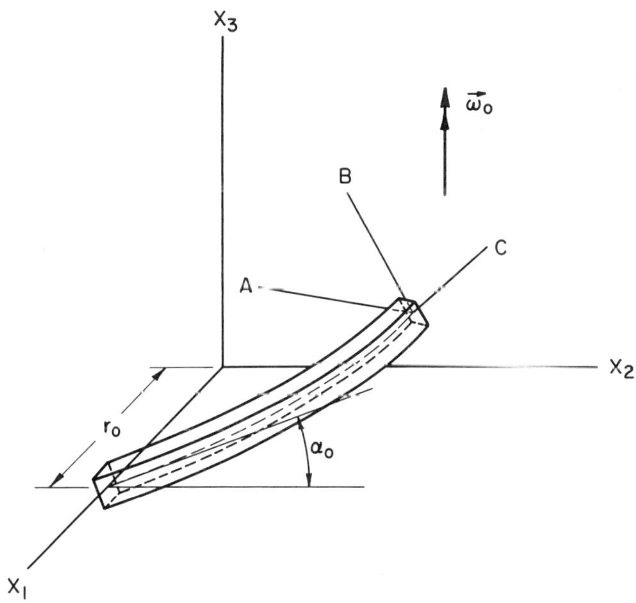

FIGURE 2.2. Undeformed helical spring with rectangular wire cross section.

Consider, for example, the helical spring under no load with the rectangular wire cross section shown in Figure 2.2. Let the angle that a tangent to the centroidal axis of the spring makes with the X_1, X_2 plane be α_0 and let the radius of the wire helix be r_0. The A-axis at any point along the centroidal axis passes through the fixed X_3-axis, as shown. If the origin of the A-, B-, and C-axes moves along the centroidal axis with a unit velocity, the angular velocity of the A, B, and C frame is

$$\vec{\omega}_0 = \frac{2\pi}{2\pi r_0/\cos\alpha_0}\vec{k} = \frac{\cos\alpha_0}{r_0}\vec{k}, \tag{2.1}$$

where $\vec{k}$ is a unit vector in the X_3 direction. Hence,

$$\kappa_0 = 0; \quad \kappa'_0 = \frac{\cos^2\alpha_0}{r_0} \quad \text{and} \quad \tau_0 = \frac{\sin\alpha_0\cos\alpha_0}{r_0}. \tag{2.2}$$

2.2 Equations of Equilibrium

Consider a thin wire loaded with the forces shown in Figure 2.3. Let s be the arc length along the wire. In Figure 2.3, N and N' are the components of the shearing force on a wire cross section in the x and y directions, respectively;

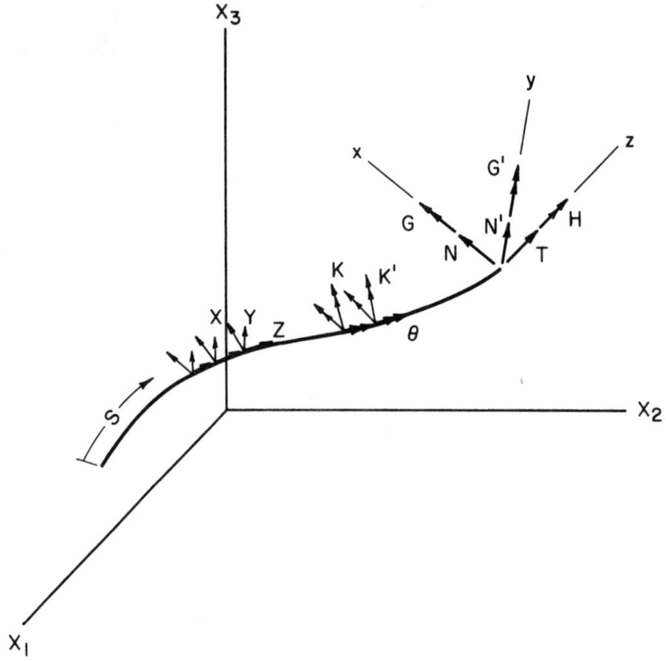

FIGURE 2.3. Loads acting on a thin wire.

T is the axial tension in the wire; G and G' are the components of the bending moment on a wire cross section in the x and y directions, respectively; H is the twisting moment in the wire; X, Y, and Z, are the components of the external line load per unit length of the centerline of the wire in the x, y, and z directions, respectively; K, K', and Θ are the components of the external moment per unit length of the centerline in the x, y, and z directions, respectively; κ and κ' are the components of curvature in the x and y directions, respectively; and τ is the twist per unit length of the wire, as defined previously.

Figure 2.4 shows a length, ds, of the centerline of the thin wire looking down the y-axis, whereas Figure 2.5 shows a similar view looking down the x-axis. For clarity only, the forces are shown (no couples) in both figures. Table 2.1 lists the direction cosines of the forces $N + dN$, $N' + dN'$, and $T + dT$ with the x-, y-, and z-axes, which are correct to within second-degree terms.

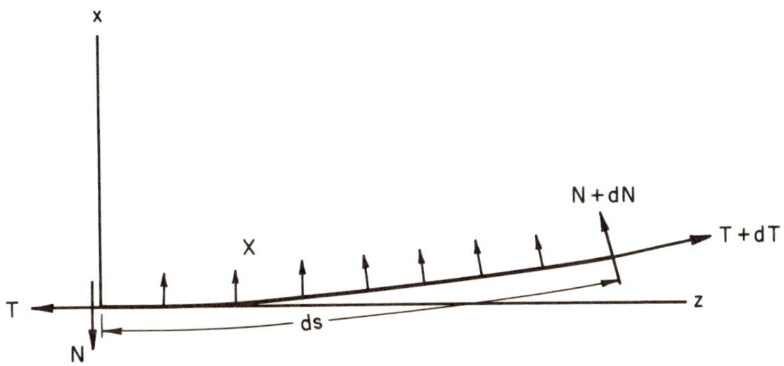

FIGURE 2.4. Centerline of a thin wire looking down the Y-axis (no couples).

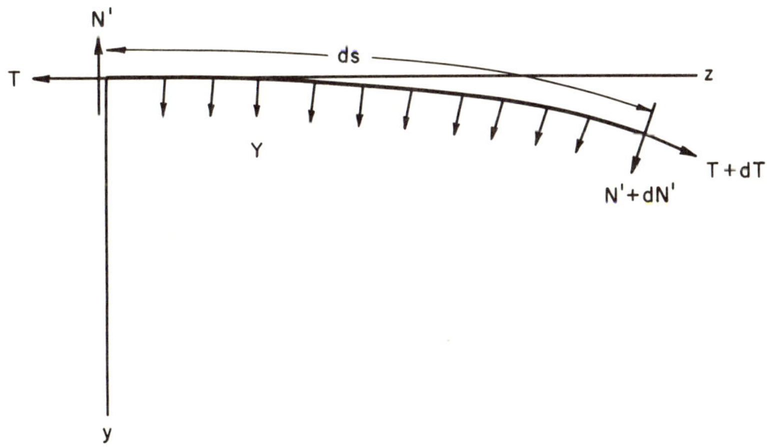

FIGURE 2.5. Centerline of a thin wire looking down the X-axis (no couples).

TABLE 2.1. Direction cosines for the forces.

Direction cosine	$N + dN$	$N' + dN'$	$T + dT$
l	1	$-\tau ds$	$\kappa' ds$
m	τds	1	$-\kappa ds$
n	$-\kappa' ds$	κds	1

A summation of forces in the x direction yields

$$X ds + dN + T\kappa' ds - N'\tau ds = 0, \tag{2.3}$$

which becomes, upon dividing by ds,

$$\frac{dN}{ds} - N'\tau + T\kappa' + X = 0. \tag{2.4}$$

Similarly, a summation of forces in the y and z directions yield

$$\frac{dN'}{ds} - T\kappa + N\tau + Y = 0 \tag{2.5}$$

and

$$\frac{dT}{ds} - N\kappa' + N'\kappa + Z = 0. \tag{2.6}$$

Figures 2.6 and 2.7 show the same element of length ds with only the couples acting on the element. The couples $G + dG$, $G' + dG'$, and $H + dH$ make the same angles with respect to the x-, y-, and z-axes as do the loads $N + dN$, $N' + dN'$, and $T + dT$, and hence, Table 2.1 can again be used. A summation of the moments about the x-axis yields

$$dG - G'\tau ds + H\kappa' ds - N' ds + K ds = 0, \tag{2.7}$$

which becomes, upon dividing by ds,

$$\frac{dG}{ds} - G'\tau + H\kappa' - N' + K = 0. \tag{2.8}$$

Similarly, a summation of moments about the y- and z-axes yield

$$\frac{dG'}{ds} - H\kappa + G\tau + N + K' = 0 \tag{2.9}$$

and

$$\frac{dH}{ds} - G\kappa' + G'\kappa + \Theta = 0. \tag{2.10}$$

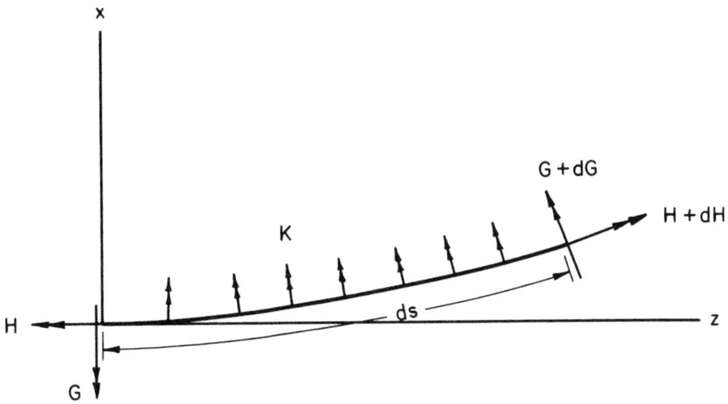

FIGURE 2.6. Centerline of a thin wire looking down the Y-axis (no forces).

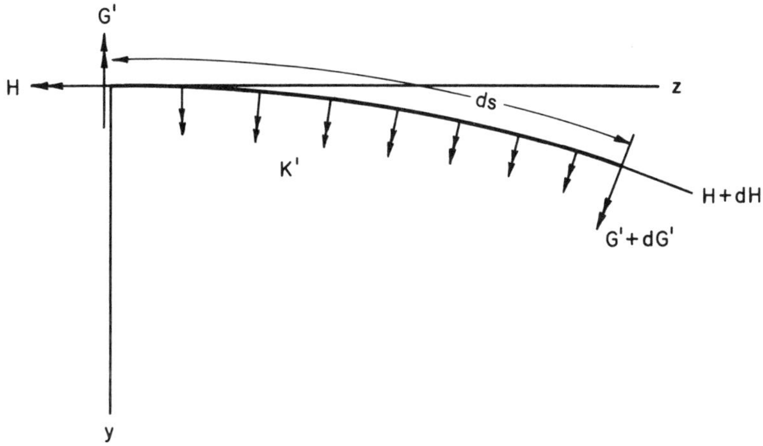

FIGURE 2.7. Centerline of a thin wire looking down the X-axis (no forces).

Equations (2.4) through (2.6) and (2.8) through (2.10) are the six differential equations of equilibrium for the thin wire loaded as shown in Figure 2.3.

2.3 Relations Between Loads and Deformations

The thin wire will now be assumed elastic with cross-sectional moments of inertia of I_x and I_y about the x- and y-axes, respectively. Also, C will denote the torsional rigidity. The expressions relating the changes in curvature and twist per unit length to the internal loads [5] are

$$G = EI_x(\kappa - \kappa_0); \quad G' = EI_y(\kappa' - \kappa'_0) \quad \text{and} \quad H = C(\tau - \tau_0), \tag{2.11}$$

where E is the modulus of elasticity of the wire material. If the wire cross section is circular, with radius R, Eq. (2.11) becomes

$$G = \frac{\pi R^4}{4} E(\kappa - \kappa_0); \quad G' = \frac{\pi R^4}{4} E(\kappa' - \kappa'_0) \quad \text{and} \quad H = \frac{\pi R^4 E}{4(1 + \nu)}(\tau - \tau_0), \tag{2.12}$$

where ν is Poisson's ratio for the wire material. The tension T in the wire is given by the expression

$$T = AE\xi, \tag{2.13}$$

where A is the cross-sectional area of the wire and ξ is the axial wire strain. For a circular cross section, Eq. (2.13) becomes

$$T = \pi R^2 E\xi. \tag{2.14}$$

The above equations will be used to generate the axial response of a simple straight strand subjected to an axial force, F, and an axial twisting moment, M_t. A simple straight strand will be defined as a strand consisting of a straight center wire of wire radius, R_1, surrounded by m_2 helical wires of wire radius, R_2.

3
Static Response of a Strand

3.1 Geometry of a Strand

Figure 3.1 shows the configuration and cross section of a loaded simple straight strand. The strand consists initially of a straight center wire of radius, R_1, surrounded by m_2 helical wires (six wires are actually shown) of wire radius, R_2. It will be assumed, for the present, that the center wire is of sufficient size to prevent the outer wires from touching each other. This is generally the case, since it tends to minimize the effect of friction in the bending of a strand. Hence, the initial radius of the helix of an outside wire is given by the expression

$$r_2 = R_1 + R_2. \tag{3.1}$$

An expression will now be derived to determine the minimum value of R_1 so that the outside wires will not be touching each other.

Consider m helical wires, in a strand, that are just touching each other. Let the radius of the helix, the wire radius, and the helix angle be denoted by r, R, and α, respectively. Figure 3.2 shows a wire cross section in a plane perpendicular to the strand. Since the wires are thin, the equation of the cross section, shown in Figure 3.2, will be assumed elliptical and, hence,

$$\left(\frac{p}{R/\sin\alpha}\right)^2 + \left(\frac{q}{R}\right)^2 = 1, \tag{3.2}$$

where (p, q) is any point on the ellipse. Now,

$$\frac{dq}{dp} = \pm \frac{p \sin^2 \alpha}{R\sqrt{1 - \left(\dfrac{p \sin\alpha}{R}\right)^2}}. \tag{3.3}$$

Also at the point (p_1, q_1), the slope is equal to $-\tan\left(\dfrac{\pi}{2} - \dfrac{\pi}{m}\right)$, as shown in

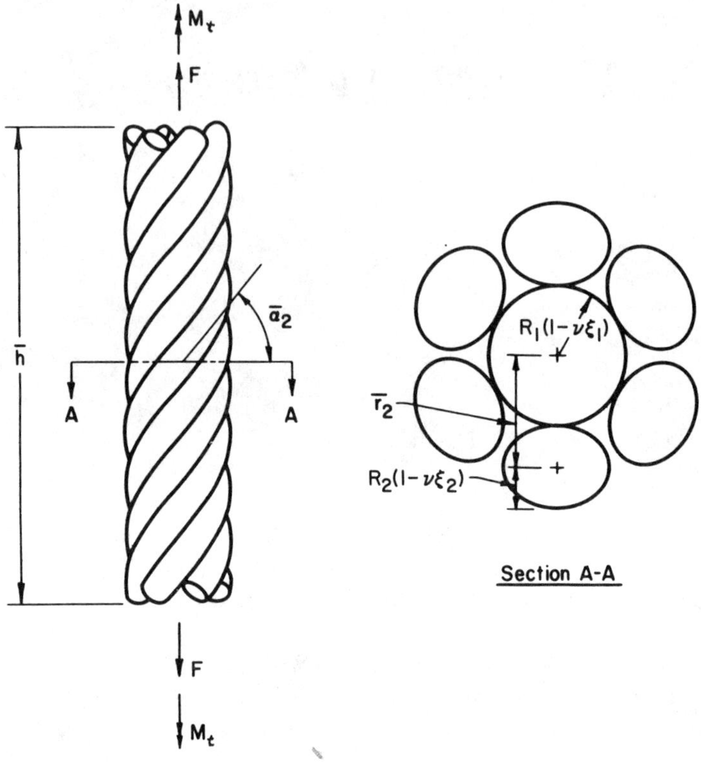

FIGURE 3.1. Loaded simple straight strand.

Figure 3.2. Hence,

$$\tan\left(\frac{\pi}{2}-\frac{\pi}{m}\right)=\frac{p_1\sin^2\alpha}{R\sqrt{1-\left(p_1\frac{\sin\alpha}{R}\right)^2}}. \tag{3.4}$$

The solution for p_1 yields

$$p_1=\frac{R}{\sin\alpha}\tan\left(\frac{\pi}{2}-\frac{\pi}{m}\right)\frac{1}{\sqrt{\sin^2\alpha+\tan^2\left(\frac{\pi}{2}-\frac{\pi}{m}\right)}}, \tag{3.5}$$

whereas Eq. (3.2) results in

$$q_1=\frac{R\sin\alpha}{\sqrt{\sin^2\alpha+\tan^2\left(\frac{\pi}{2}-\frac{\pi}{m}\right)}}. \tag{3.6}$$

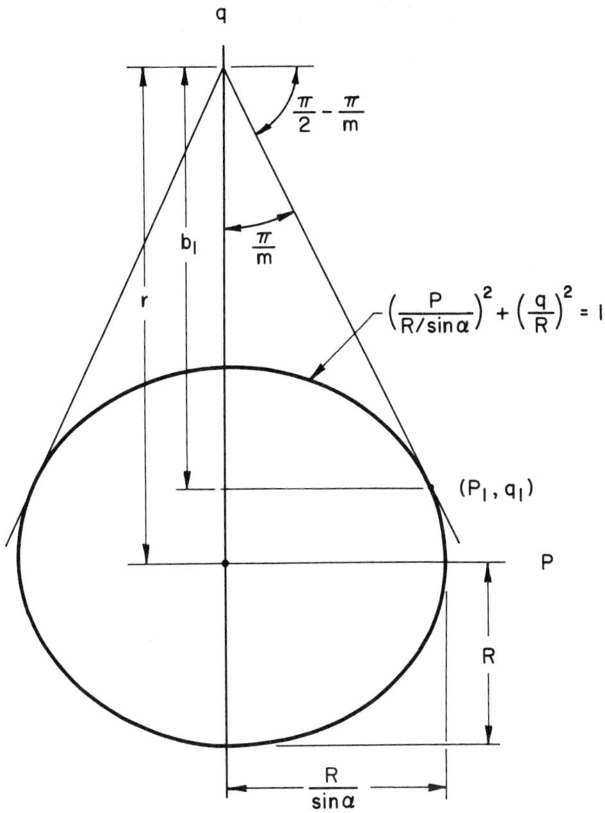

FIGURE 3.2. Wire cross section perpendicular to axis of strand.

Figure 3.2 indicates that

$$b_1 = p_1 \tan\left(\frac{\pi}{2} - \frac{\pi}{m}\right) \tag{3.7}$$

and, hence, since $r = b_1 + q_1$,

$$r = R\sqrt{1 + \frac{\tan^2\left(\frac{\pi}{2} - \frac{\pi}{m}\right)}{\sin^2\alpha}}. \tag{3.8}$$

Equation (3.8) yields the radius of the wire helix in which the wires are just touching each other. Hence, in the simple straight strand,

$$R_2\sqrt{1 + \frac{\tan^2\left(\frac{\pi}{2} - \frac{\pi}{m_2}\right)}{\sin^2\alpha}} < R_1 + R_2 \tag{3.9}$$

if the outside wires are not to touch each other.

3.2 Axial Response of a Simple Straight Strand

Consider again the configuration and cross section of a loaded simple straight strand shown in Figure 3.1. The initial helix angle α_2 of an outside wire is determined by the relation

$$\tan \alpha_2 = \frac{p_2}{2\pi r_2}, \tag{3.10}$$

where p_2 is the initial pitch of an outside wire. The original components of the curvature and the twist per unit length are [see Eq. (2.2)]

$$\kappa_2 = 0; \quad \kappa'_2 = \frac{\cos^2 \alpha_2}{r_2} \quad \text{and} \quad \tau_2 = \frac{\sin \alpha_2 \cos \alpha_2}{r_2}. \tag{3.11}$$

Let the wires in the strand now be deformed under the action of the axial loads F, and M_t, where F is the total axial force and M_t is the total axial twisting moment. The outside wire, under the loading, will assume to be deformed into a new helix, where

$$\bar{\kappa}_2 = 0; \quad \bar{\kappa}'_2 = \frac{\cos^2 \bar{\alpha}_2}{\bar{r}_2} \quad \text{and} \quad \bar{\tau}_2 = \frac{\sin \bar{\alpha}_2 \cos \bar{\alpha}_2}{\bar{r}_2}. \tag{3.12}$$

The barred symbols refer to the previously defined quantities in the deformed state or final state.

It will now be assumed that an outside wire is not subjected to external bending moments per unit length, that is, $K_2 = K'_2 = 0$, and that the axial wire tension T_2 is constant along the length of the wire. Then by virtue of Eqs. (2.12), (3.11) and (3.12), the equations of equilibrium [Eqs. (2.4) through (2.6) and (2.8) through (2.10)] become

$$-N'_2\bar{\tau}_2 + T_2\bar{\kappa}'_2 + X_2 = 0, \tag{3.13}$$

$$Y_2 = 0, \tag{3.14}$$

$$Z_2 = 0, \tag{3.15}$$

$$-G'_2\bar{\tau}_2 + H_2\bar{\kappa}'_2 - N'_2 = 0, \tag{3.16}$$

$$N_2 = 0, \tag{3.17}$$

and

$$\Theta_2 = 0. \tag{3.18}$$

The subscript 2 refers to the outside wires. Figure 3.3 shows the loads acting on an helical wire; the positive directions of the loads are shown.

Equations (3.13) and (3.16) can be regarded as determining the values of X_2 and N'_2 required to hold an outside helical wire in equilibrium for given values of $\bar{\alpha}_2, \bar{r}_2$, and T_2. It should be noted that the equations of equilibrium and Eq. (2.12) constitute a set of nonlinear equations and are valid for large deflections. Large deflections could occur, for instance, in the case of a thin wire helical

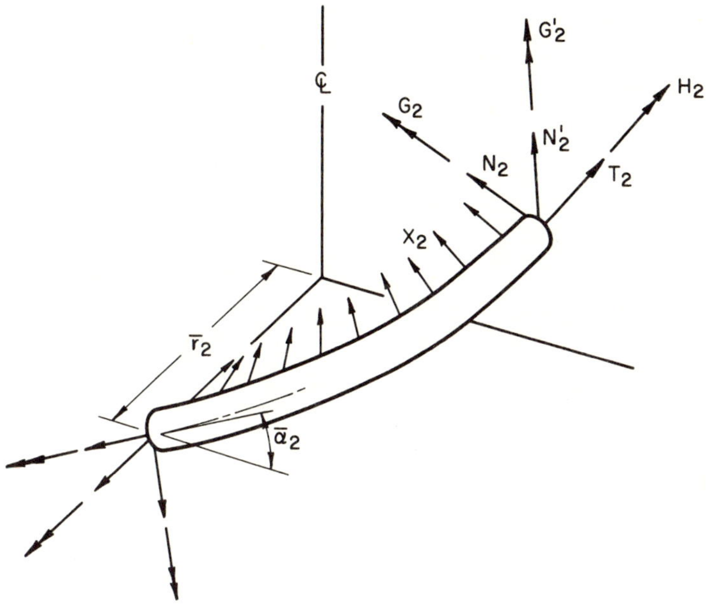

FIGURE 3.3. Loads acting on helical wire.

spring in which the value of α_2 is generally small. In the case of wire rope, however, the value of α_2 is generally large and the change in α_2, $\Delta\alpha_2$, where

$$\Delta\alpha_2 = \bar{\alpha}_2 - \alpha_2 \tag{3.19}$$

is small. Use will be made of this fact later on in simplifying the solution.

The axial strain ε of a straight strand is defined as

$$\varepsilon = \frac{\bar{h} - h}{h}, \tag{3.20}$$

where h is the original length of the strand and $\bar{h}$ is the final length of the strand. Figure 3.4, a developed view of the centerline of an outer wire, indicates the length h and $\bar{h}$. The rotational strain β_2 of an outer wire will be defined as

$$\beta_2 = r_2 \frac{(\bar{\theta}_2 - \theta_2)}{h}, \tag{3.21}$$

where θ_2 and $\bar{\theta}_2$ are the initial and final angle, respectively, that an outer wire sweeps out in a plane perpendicular to the axis of the strand. The angle of twist per unit length, τ_s, of the strand is defined by the expression

$$\tau_s = \frac{(\bar{\theta}_2 - \theta_2)}{h}, \tag{3.22}$$

An analysis of the configuration shown in Figure 3.4 yields

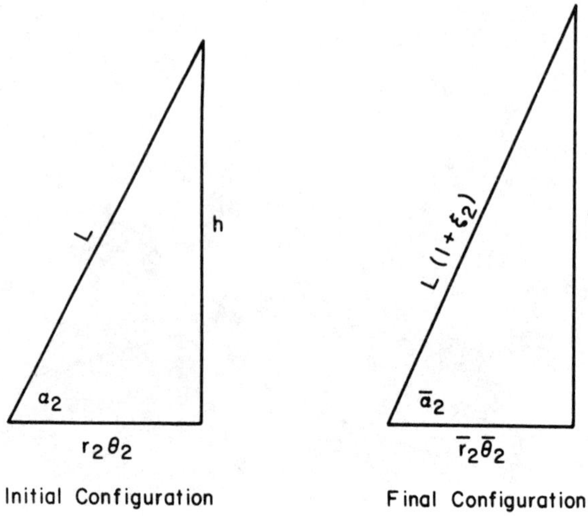

FIGURE 3.4. Developed view of outer wire centerline.

$$\varepsilon = \frac{\bar{h} - h}{h} = \xi_1 = (1 + \xi_2)\frac{\sin \bar{\alpha}_2}{\sin \alpha_2} - 1 \tag{3.23}$$

and

$$\beta_2 = r_2\tau_s = \frac{r_2}{\bar{r}_2}\frac{(1 + \xi_1)}{\tan \bar{\alpha}_2} - \frac{1}{\tan \alpha_2}, \tag{3.24}$$

where ξ_1 is the axial strain in the center wire ($\xi_1 = \varepsilon$), and ξ_2 is the axial strain in an outer wire.

Let

$$|\Delta\alpha_2| = |\bar{\alpha}_2 - \alpha_2| \ll 1, \tag{3.25}$$

which is valid for most metallic strands. Hence, $\sin \bar{\alpha}_2$ can be expressed as

$$\sin \bar{\alpha}_2 = \sin(\alpha_2 + \Delta\alpha_2) = \sin \alpha_2 + \Delta\alpha_2 \cos \alpha_2, \tag{3.26}$$

where higher-ordered terms are neglected. Equation (3.23) can now be written as

$$\xi_1 = \xi_2 + \frac{\Delta\alpha_2}{\tan \alpha_2} = \varepsilon, \tag{3.27}$$

where ξ_1 and ξ_2 are assumed small. Equation (3.24), after a similar procedure, becomes

$$\beta_2 = \frac{r_2}{\bar{r}_2}\left[\frac{1 + \xi_2}{\tan \alpha_2} - \Delta\alpha_2\right] - \frac{1}{\tan \alpha_2}. \tag{3.28}$$

The final helical radius $\bar{r}_2$ becomes, due to Poisson's ratio effect,

$$\bar{r}_2 = R_1(1 - \nu\xi_1) + R_2(1 - \nu\xi_2), \tag{3.29}$$

where the contact deformation in the center and the outer wire is neglected. Now $r_2/\bar{r}_2$ can be expressed as

$$\frac{r_2}{\bar{r}_2} = \frac{R_1 + R_2}{R_1 + R_2 - \nu(R_1\xi_1 + R_2\xi_2)} = 1 + \nu\frac{(R_1\xi_1 + R_2\xi_2)}{r_2}, \tag{3.30}$$

since ξ_1 and ξ_2 are small. Hence Eq. (3.28) becomes

$$\beta_2 = r_2\tau_s = \frac{\xi_2}{\tan\alpha_2} - \Delta\alpha_2 + \nu\frac{(R_1\xi_1 + R_2\xi_2)}{r_2\tan\alpha_2}. \tag{3.31}$$

The change in curvature $\Delta\kappa'_2$ and the change in twist per unit length $\Delta\tau_2$ can also be linearized. The following results:

$$\begin{aligned} R_2\Delta\kappa'_2 &= \frac{\cos^2\bar{\alpha}_2}{\bar{r}_2/R_2} - \frac{\cos^2\alpha_2}{r_2/R_2} \\ &= -\frac{2\sin\alpha_2\cos\alpha_2}{r_2/R_2}\Delta\alpha_2 + \nu\frac{(R_1\xi_1 + R_2\xi_2)}{r_2}\frac{\cos^2\alpha_2}{r_2/R_2} \end{aligned} \tag{3.32}$$

and

$$\begin{aligned} R_2\Delta\tau_2 &= \frac{\sin\bar{\alpha}_2\cos\bar{\alpha}_2}{\bar{r}_2/R_2} - \frac{\sin\alpha_2\cos\alpha_2}{r_2/R_2} \\ &= \frac{(1 - 2\sin^2\alpha_2)}{r_2/R_2}\Delta\alpha_2 + \nu\frac{(R_1\xi_1 + R_2\xi_2)}{r_2}\frac{\sin\alpha_2\cos\alpha_2}{r_2/R_2}. \end{aligned} \tag{3.33}$$

The following can now be written down for an outside wire:

$$\frac{G'_2}{ER_2^3} = \frac{\pi}{4}R_2\Delta\kappa'_2, \tag{3.34}$$

$$\frac{H_2}{ER_2^3} = \frac{\pi}{4(1+\nu)}R_2\Delta\tau_2, \tag{3.35}$$

$$\frac{N'_2}{ER_2^2} = \frac{H_2}{ER_2^3}\frac{\cos^2\alpha_2}{r_2/R_2} - \frac{G'_2}{ER_2^3}\frac{\sin\alpha_2\cos\alpha_2}{r_2/R_2}, \tag{3.36}$$

$$\frac{T_2}{ER_2^2} = \pi\xi_2, \tag{3.37}$$

and

$$\frac{X_2}{ER_2} = \frac{N'_2}{ER_2^2}\frac{\sin\alpha_2\cos\alpha_2}{r_2/R_2} - \frac{T_2}{ER_2^2}\frac{\cos^2\alpha_2}{r_2/R_2}. \tag{3.38}$$

Equations (3.36) and (3.38) are obtained from the equations of equilibrium [Eqs. (3.16) and (3.13)], where it assumed that displacements are small.

A projection of the forces, acting on the outside wires, in the axial direction of the strand yields

$$\frac{F_2}{ER_2^2} = m_2\left[\frac{T_2}{ER_2^2}\sin\alpha_2 + \frac{N'_2}{ER_2^2}\cos\alpha_2\right], \tag{3.39}$$

where F_2 is the total axial force in the strand acting on the m_2 outer wires. The total axial twisting moment M_2 acting on the outside wires is

$$\frac{M_2}{ER_2^3} = m_2\left[\frac{H_2}{ER_2^3}\sin\alpha_2 + \frac{G'_2}{ER_2^3}\cos\alpha_2 + \frac{T_2}{ER_2^2}\frac{r_2}{R_2}\cos\alpha_2 - \frac{N'_2}{ER_2^2}\frac{r_2}{R_2}\sin\alpha_2\right]. \tag{3.40}$$

The axial force F_1 and the axial twisting moment M_1 acting on the center wire are given by the expressions

$$\frac{F_1}{ER_1^2} = \pi\xi_1 \tag{3.41}$$

and

$$\frac{M_1}{ER_1^3} = \frac{\pi}{4(1+\nu)}R_1\tau_s. \tag{3.42}$$

The total axial force F and the total axial twisting moment M_t acting on the stand can be written as

$$F = F_1 + F_2 \tag{3.43}$$

and

$$M_t = M_1 + M_2. \tag{3.44}$$

The contact force per unit length X_2 becomes an internal force as far as the strand is concerned. Once this force is known, an estimate can be made of the contact stresses [6].

The above equations will be used in the example worked out below.

Example 3.1. Consider a simple straight strand with $R_1 = 0.103$ in., $R_2 = 0.101$ in., $p_2 = 9.75$ in., $E = 28{,}500{,}000$ psi, $\nu = 0.25$, and $m_2 = 6$. Equation (3.1) yields (assuming that the outside wires are not touching each other)

$$r_2 = 0.103 + 0.101 = 0.204 \text{ in.}$$

The helix angle α_2 is determined by Eq. (3.10); hence,

$$\tan\alpha_2 = \frac{9.75}{2\pi \times 0.204} = 7.60667,$$

and, therefore, $\alpha_2 = 82.5106°$. The following values can now be written down: $\sin\alpha_2 = 0.99147$, $\sin^2\alpha_2 = 0.98301$, $\cos\alpha_2 = 0.13034$, $\cos^2\alpha_2 = 0.016989$,

$\sin \alpha_2 \cos \alpha_2 = 0.12923$, and $r_2/R_2 = 2.01980$. The outside wires should be checked to determine if they are touching each other. Since there are six outsdide wires, Eq. (3.9) yields

$$R_2 \sqrt{1 + \frac{\tan^2\left(\frac{\pi}{2} - \frac{\pi}{m_2}\right)}{\sin^2 \alpha_2}} = 0.2033 \text{ in.} < 0.204 \text{ in.} = R_1 + R_2,$$

and so the outside wires do not touch each other.

Now let the angle of twist per unit length of the strand $\tau_s = 0$ (the strand is not allowed to rotate) and let $\xi_1 = \varepsilon = 0.003$. Equations (3.27) and (3.31) become

$$0.003 = \xi_2 + \frac{\Delta\alpha_2}{7.60667}$$

and

$$0 = \frac{\xi_2}{7.60667} - \Delta\alpha_2 + 0.25\frac{(0.103 \times 0.003 + 0.101\xi_2)}{0.204 \times 7.60667},$$

which have the solution $\xi_2 = 0.002936$ and $\Delta\alpha_2 = 0.00048359$. The values of $R_2\Delta\kappa'_2$ and $R_2\Delta\tau_2$ can be computed from Eqs. (3.32) and (3.33). The following results:

$$\begin{aligned} R_2\Delta\kappa'_2 &= -\frac{2 \times 0.12923 \times 0.00048359}{2.01980} \\ &\quad + \frac{0.25(0.103 \times 0.003 + 0.101 \times 0.002963) \times 0.016989}{0.204 \times 2.0198} \\ &= -0.00005564 \end{aligned}$$

and

$$\begin{aligned} R_2\Delta\tau_2 &= \frac{(1 - 2 \times 0.98301) \times 0.00048359}{2.01980} \\ &\quad + \frac{0.25(0.103 \times 0.003 + 0.101 \times 0.002936) \times 0.12923}{0.204 \times 2.0198} \\ &= -0.0001838 \end{aligned}$$

are obtained. Eqs. (3.34) through (3.44) yield

$$\frac{G'_2}{ER_2^3} = -0.00004370,$$

$$\frac{H_2}{ER_2^3} = -0.00011549,$$

$$\frac{N'_2}{ER_2^2} = 0.000001825,$$

$$\frac{T_2}{ER_2^2} = 0.00922505,$$

$$\frac{X_2}{ER_2} = -0.000077474,$$

$$\frac{F_2}{ER_2^2} = 0.054879,$$

$$\frac{M_2}{ER_2^3} = 0.013828,$$

$$\frac{F_1}{ER_1^2} = 0.00942478,$$

$$\frac{M_1}{ER_1^3} = 0,$$

where $F = 2850 + 15{,}955 = 18{,}805$ lb and $M_t = 0 + 406 = 406$ in. lb.

It is interesting to note that the inside and outside wires carry 15.2% and 84.8% of the total axial load. Also, since the above equations are linear, a reduction or increase in the axial strain, with $\tau_s = 0$, would correspond to a similar decrease or increase in the loads. For example, with $\varepsilon = 0.0015$ and $\tau_s = 0$, the total axial force would be 9,400 lb and the total axial moment would be 203 in lb.

3.3 Stress Determination of a Simple Straight Strand

In the previous section, the loads acting on the individual wires for the case of an axially loaded simple straight strand were determined. In this section the stresses caused by these loads are investigated. It will be assumed, at this point, that the wires are initially stress free.

In the case of the center wire, the axial wire stress is

$$_F\sigma_1 = \frac{F_1}{\pi R_1^2}, \tag{3.45}$$

whereas the maximum shearing stress on the cross section is

$$_M\sigma_1 = \frac{2M_1}{\pi R_1^3}. \tag{3.46}$$

The outside wires are subjected to axial, bending, and torsional loadings in addition to the shearing load N'_2. The stresses caused by the shearing force N'_2 are in general very small and will be neglected. The axial stress caused by the load T_2 is

$$ {}_T\sigma_2 = \frac{T_2}{\pi R_2^2}, \tag{3.47}$$

whereas the maximum normal stress due to the bending moment G'_2 is

$$ {}_{G'}\sigma_2 = \frac{4G'_2}{\pi R_2^3}. \tag{3.48}$$

The maximum shearing stress on an outside wire due to the twisting moment H_2 is

$$ {}_H\sigma_2 = \frac{2H_2}{\pi R_2^3}. \tag{3.49}$$

Example 3.2. Consider the strand used in Example 3.1. Let the strand be subjected to an axial load of 18,805 lb and not allowed to rotate ($\tau_s = 0$). The results of Example 3.1 and Eqs. (3.45) through (3.49) yield

$$ {}_F\sigma_1 = \frac{2850}{\pi(0.103)^2} = 85{,}500 \text{ psi}$$

$$ {}_M\sigma_1 = 0 \text{ psi}$$

$$ {}_T\sigma_2 = \frac{2682}{\pi(0.101)^2} = 83{,}700 \text{ psi}$$

$$ {}_G\sigma_2 = \frac{4 \times 1.283}{\pi(0.101)^3} = 1{,}580 \text{ psi}$$

and

$$ {}_H\sigma_2 = \frac{2 \times 3.391}{\pi(0.101)^3} = 2{,}090 \text{ psi}.$$

The shearing force N'_2 is 0.53 lb. The maximum normal tensile stress acting on an outer wire is

$$ {}_T\sigma_2 + {}_{G'}\sigma_2 = 83{,}700 + 1{,}580 = 85{,}280 \text{ psi},$$

and this stress occurs on the inside of an outer wire (due to the sign of G'_2).

It should be noted that the center wire suffers a slightly greater stress than the outer wire. Again, the remark can be made that, an increase or a decrease in the axial load (with $\tau_s = 0$) would cause a similar increase or decrease in the stresses. The contact stresses are, of course, an exception to this, since they do not depend linearly on the loads. Contact stresses will be considered later.

3.4 Load Deformation Relation for a Simple Straight Strand

In general, the total axial force F and the total axial twisting moment M acting on a strand can be expressed as

$$\frac{F}{AE} = C_1\varepsilon + C_2\beta \tag{3.50}$$

and

$$\frac{M_t}{ER^3} = C_3\varepsilon + C_4\beta, \tag{3.51}$$

where

$$A = \Sigma\pi R_i^2. \tag{3.52}$$

A is the total metallic area of the strand, R_i is the radius of an individual wire, $C_1, \ldots, C_4$ are constants, which can be determined analytically, R is the radius of the strand, ε is the axial strain, and β is the rotational strain of the strand defined by the equation

$$\beta = R\tau_s, \tag{3.53}$$

where τ_s is the angle of twist per unit length of the strand.

As mentioned previously, the value of the constants in Eqs. (3.50) and (3.51) can be determined analytically for a given strand. Letting, for example, $\beta = R\tau_s = 0$ and $\varepsilon = \xi_1$ equals a given value, Eqs. (3.43) and (3.44) can be used to determine F and M_t. Hence, with F, M_t, ε, and β known, Eqs. (3.50) and (3.51) yield C_1 and C_3. Now, let $\varepsilon = 0$ and β equals a given value. Again, F and M_t can be calculated and Eqs. (3.50) and (3.51) will yield C_2 and C_4.

It is instructive at this point to define the effective modulus E_e of a strand by the relation

$$E_e = C_1 E, \tag{3.54}$$

which is the modulus of the strand when $R\tau_s = \beta = 0$, that is,

$$\frac{F}{A} = C_1 E\varepsilon = E_e\varepsilon. \tag{3.55}$$

Example 3.3. Consider again the strand used in Examples 3.1 and 3.2. The metallic area is

$$A = \pi(0.103)^2 + 6 \times \pi(0.101)^2 = 0.2256 \text{ in.}^2$$

and

$$R = 0.103 + 2 \times 0.101 = 0.305 \text{ in.}$$

Equations (3.50) and (3.51) and the results of Example 3.1 yield

$$C_1 = \frac{18{,}805}{0.2256 \times 28{,}500{,}000 \times 0.003} = 0.975$$

and

$$C_3 = \frac{406}{28{,}500{,}000 \times (0.305)^3 \times 0.003} = 0.167.$$

Now let $\xi_1 = \varepsilon = 0$ and $\xi_2 = 0.001$. Equation (3.27) yields

$$0 = 0.001 + \frac{\Delta\alpha_2}{7.6067}$$

and, hence $\Delta\alpha_2 = -0.0076067$. Equation (3.31) yields

$$0.204\tau_s = \frac{0.001}{7.6067} + 0.0076067 + \frac{0.25(0.101 \times 0.001)}{0.204 \times 7.6067}$$

and, therefore, $\tau_s = 0.038012$, with $\beta = 0.305 \times 0.038012 = 0.011593$. Equations (3.32) through (3.44) result in

$$R_2\Delta K'_2 = \frac{2 \times 0.12923 \times 0.0076067}{2.01980} + \frac{0.25(0.101 \times 0.001) \times 0.016989}{0.204 \times 2.1980} = 0.00097442,$$

$$R_2\Delta\tau_s = -\frac{(1 - 2 \times 0.98301) \times 0.0076067}{2.01980} + \frac{0.25(0.101 \times 0.001) \times 0.129230}{0.204 \times 2.01980} = 0.0036460,$$

$$\frac{G'_2}{ER_2^3} = 0.0007653,$$

$$\frac{H_2}{ER_2^3} = 0.0022908,$$

$$\frac{N_2'}{ER_2^2} = 0.00002970,$$

$$\frac{T_2}{ER_2^2} = 0.0031416,$$

$$\frac{X_2}{ER_2} = -0.00002832,$$

$$\frac{F_2}{ER_2^2} = 0.018665,$$

$$\frac{M_2}{ER_2^3} = 0.019546,$$

$$\frac{F_1}{ER_1^2} = 0,$$

$$\frac{M_1}{ER_1^3} = 0.00246,$$

where $F = 5{,}426$ lb and $M_t = 76.6 + 573.9 = 650.5$ in. lb.
Equations (3.50) and (3.51) yield

$$C_2 = \frac{5426}{0.2256 \times 28{,}500{,}000 \times 0.011593} = 0.0728$$

and

$$C_4 = \frac{650.5}{28{,}500{,}000 \times (0.305)^3 \times 0.011593} = 0.0694.$$

Therefore, Eqs. (3.50) and (3.51) become, for the given strand,

$$\frac{F}{AE} = 0.975\varepsilon + 0.0728\beta$$

and

$$\frac{M_t}{ER^3} = 0.167\varepsilon + 0.0694\beta.$$

It should be noted that the constants C_1, C_2, C_3, and C_4 are independent of the modulus of elasticity, E and depend only on α_2, R_2/R_1, and ν.

3.5 Pure Bending of a Simple Straight Strand

Consider an initially straight helical spring with a helix angle, α, and let the spring be subjected to a bending moment, m_s, which is applied about an axis perpendicular to the original axis of the spring. Figure 3.5 shows such a spring.

Since the initial configuration of the wire is a helical spring, the initial curvatures and twist per unit length are

$$K = 0; \quad K' = \frac{\cos^2 \alpha}{r} \quad \text{and} \quad \tau = \frac{\sin \alpha \cos \alpha}{r}, \tag{3.56}$$

where r is the initial radius of the helix and, since the spring is subjected to a pure bending moment only, the following results:

$$X = Y = Z = K = K' = \Theta = N = N' = T = 0. \tag{3.57}$$

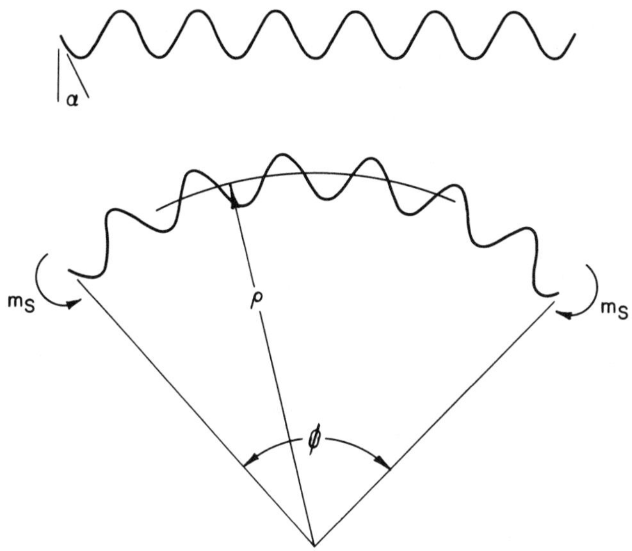

FIGURE 3.5. Helical spring bent by couple.

The equations of equilibrium Eqs. (2.4) through (2.6) and (2.8) through (2.10) yield

$$\frac{dG}{ds} - G'\tau_1 + H\kappa_1' = 0, \tag{3.58}$$

$$\frac{dG'}{ds} - H\kappa_1 + G\tau_1 = 0, \tag{3.59}$$

and

$$\frac{dH}{ds} - G\kappa_1' + G'\kappa_1 = 0, \tag{3.60}$$

where κ_1, κ_1', and τ_1 are the final (deformed) curvatures and twist per unit length. Then by virtue of Eqs. (2.12) and (3.56), the equilibrium equations [Eqs. (3.58) through (3.60)] can be written as

$$\frac{dG}{ds} - \frac{4}{\pi R^4 E}\nu G'H - \frac{\sin\alpha\cos\alpha}{r}G' + \frac{\cos^2\alpha}{r}H = 0, \tag{3.61}$$

$$\frac{dG'}{ds} + \frac{4}{\pi R^4 E}\nu GH + \frac{\sin\alpha\cos\alpha}{r}G = 0, \tag{3.62}$$

and

$$\frac{dH}{ds} - \frac{\cos^2\alpha}{r}G = 0. \tag{3.63}$$

Equations (3.61) through (3.63) constitute a nonlinear system of first-order ordinary differential equations that can be integrated numerically (Euler's method or Picard's method, for example) under suitable initial conditions. If these three equations are, respectively, multiplied by G, G', and H and the resulting equations are added, the following results:

$$\frac{1}{2}\frac{d}{ds}(G^2 + G'^2 + H^2) = 0, \tag{3.64}$$

which states that the magnitude of the resulting moment on any cross section is constant (independent of s).

One Eqs. (3.61) through (3.63) are integrated numerically for G, G', and H, Eq. (2.12) can be used to calculate the final curvature and twist. When the curvatures and twist are known, the deformation of the centerline of the deformed wire is completely determined apart from its position in space [7].

In most engineering problems, however, an exact determination of the centerline of the deformed wire is not needed. A more practical approach would be to assume that, under the action of a pure bending moment, m_s, applied perpendicular to the original axis of the helix, the spring behaves like a beam, as shown in Figure 3.1. The initially straight axis of the spring then deforms into a circle of radius ρ with the angle ϕ shown.

Consider, for example, the case of $v = 0$. Equations (3.61) through (3.63) become linear and therefore have the solution

$$G = C_1 \cos ks + C_2 \sin ks \tag{3.65}$$

$$H = C_1 \cos\alpha \sin ks - C_2 \cos\alpha \cos ks + C_3 \tag{3.66}$$

and

$$G' = -C_1 \sin\alpha \sin ks + C_2 \sin\alpha \cos ks + \frac{\cos\alpha}{\sin\alpha} C_3, \tag{3.67}$$

where

$$k = \frac{\cos\alpha}{r}. \tag{3.68}$$

Let $G = m_s$, $H = 0$, $G' = 0$ at $s = 0$. This yields

$$G = m_s \cos ks; \quad H = m_s \cos\alpha \sin ks; \quad G' = -m_s \sin\alpha \sin ks. \tag{3.69}$$

The strain energy in the spring U can be written [5] as

$$\begin{aligned} U &= \frac{1}{2}\int_0^l \left[\frac{\pi R^4 E}{4}(\kappa_1 - \kappa)^2 + \frac{\pi R^4 E}{4}(\kappa'_1 - \kappa')^2 + \frac{\pi R^4 E}{4}(\tau_1 - \tau)^2\right] ds \\ &= \frac{1}{2}\int_0^l \frac{4}{\pi R^4 E}[G^2 + G'^2 + H^2] ds = \frac{2}{\pi R^4 E}\int_0^l m_s^2 ds = \frac{2m_s^2 l}{\pi R^4 E}, \end{aligned} \tag{3.70}$$

where l is the length of the wire.

When the work done by the bending moment m_s is equated to the strain energy, the result is

$$\int_0^{\phi} m_s(\phi)d\phi = \frac{2m_s^2 l}{\pi R^4 E}. \tag{3.71}$$

A differentiation of Eq. (3.71) yields

$$m_s = \frac{4m_s l}{\pi R^4 E}\frac{dm_s}{d\phi}, \tag{3.72}$$

whereas an integration of Eq. (3.72) yields (since $m_s(0) = 0$)

$$m_s = \frac{\pi E R^4}{4l}\phi. \tag{3.73}$$

Now let

$$h = l \sin \alpha, \tag{3.74}$$

where h is the length of the spring. Then Eq. (3.73) becomes

$$\frac{4m_s}{\pi E R^4 \sin \alpha} = \frac{\phi}{h} = \frac{1}{\rho}. \tag{3.75}$$

The above equation is valid for $v = 0$.

When the exact solution for $v = 0$ is used in Picard's method to obtain a solution when $v \neq 0$ [8], the following results:

$$\frac{1}{\rho} = \frac{(2 + v\cos^2 \alpha)}{2 \sin \alpha}\frac{4m_s}{\pi R^4 E}. \tag{3.76}$$

Equation (3.76) is valid for large changes in curvature. It should be noted that as α approaches 90°, the curvature $1/\rho$ approaches that of a straight beam.

Consider now a simple straight strand bent into a circle of radius, ρ, by a bending moment, M_b. In this case, friction will be neglected and the bending stiffness A^* of the straight strand will be approximated by the bending stiffness of each wire in the strand, that is, the strand is treated as an assemblages of helical springs [9]. Recent experimental investigations tend to confirm this [10]. This approximation is felt to be a reasonable one for a simple strand, since the outside wires are not touching each other; as a result, when the strand is bent, the outside wires act independently of each other. Since friction is neglected, the center wire acts independently of the outer wires. In Chapter 5, it will be noted that friction plays a small role in determining the bending stiffness of a simple strand.

On the basis of the previous discussion, the following expression can be written:

$$M_b = \frac{\pi E}{4}\left[\frac{2m_2 \sin \alpha_2}{(2 + v\cos^2 \alpha_2)}R_2^4 + R_1^4\right]\frac{1}{\rho} = \frac{A^*}{\rho}, \tag{3.77}$$

where M_b is the total bending moment applied to the strand, ρ is the radius of curvature of the strand, and A^* is the bending stiffness, of the simple strand, defined by the equation

$$A^* = \frac{\pi E}{4}\left[\frac{2m_2 \sin\alpha_2}{(2 + \nu\cos^2\alpha_2)} R_2^4 + R_1^4\right]. \tag{3.78}$$

Example 3.4. Consider again the strand used in Example 3.1. The bending stiffness of this strand is given by Eq. (3.78) and, hence,

$$\begin{aligned} A^* &= \frac{\pi \times 28{,}500{,}000}{4}\left[\frac{6 \times 2 \times 0.99147 \times (0.101)^4}{(2 + 0.25 \times 0.016989)} + (0.103)^4\right] \\ &= 16{,}350 \text{ lb in.}^2. \end{aligned}$$

Let, for example, $\rho = 8$ ft. In this case, Eq. (3.76) yields

$$m_s = \frac{\pi \times 28{,}500{,}000 \times 2 \times 0.99147 \times (0.101)^4}{4(2 + 0.25 \times 0.016989) \times 8 \times 12} = 24.0 \text{ in. lb},$$

and Eq. (3.77) yields

$$M_b = \frac{16{,}350}{8 \times 12} = 170.3 \text{ in. lb.}$$

Notice the small effect of ν on the value of the bending stiffness A^*.

3.6 Stress Determination of a Strand Subjected to Bending

Equation (3.69) determines G_2, G'_2, and H_2, the components of the bending moment m_s in the outer wires for the case of $\nu = 0$. Based on the previous equations, this expression is felt to be a good approximation for the case of $\nu \neq 0$ and therefore will be used to determine the stresses in a outer wire in a simple strand subjected to bending. Hence,

$${}_G\sigma_2 = \frac{4m_s}{\pi R_2^3}\cos k_2 s, \tag{3.79}$$

$${}_{G'}\sigma_2 = \frac{4m_s}{\pi R_2^3}\sin\alpha_2 \sin k_2 s, \tag{3.80}$$

and

$${}_H\sigma_2 = \frac{2m_s}{\pi R_2^3}\cos\alpha_2 \sin k_2 s, \tag{3.81}$$

where

$$k_2 = \frac{\cos \alpha_2}{r_2}, \tag{3.82}$$

$$m_s = \frac{\pi R_2^4 E \sin \alpha_2}{2\rho(2 + \nu \cos^2 \alpha_2)}, \tag{3.83}$$

${}_G\sigma_2$ and ${}_{G'}\sigma_2$ are the maximum normal bending stresses on a given cross section due to the bending moments G_2 and G'_2, ${}_H\sigma_2$ is the maximum shear stress on a given cross section due to the twisting moment H_2. The maximum normal stress on the cross section ${}_m\sigma_2$ occurs at $s = 0$, $s = \pi(r_2/\cos \alpha_2), \ldots,$ and, therefore,

$${}_m\sigma_2 = \frac{4m_s}{\pi R_2^3}. \tag{3.84}$$

The center wire is also subjected to pure bending and, therefore, the maximum bending stress in the center wire ${}_m\sigma_1$ is

$${}_m\sigma_1 = \frac{ER_1}{\rho}. \tag{3.85}$$

The maximum bending stress will always occur in the center wire for two reasons: (1) the center wire has a larger wire radius than the outside wire, that is $R_1 > R_2$ and (2) the helix angle α_2 tends to decrease the stiffness of an outside wire, compared with a straight wire. For example, a helical spring has a smaller bending stiffness, compared with a straight wire of the same wire diameter.

Example 3.5. Consider the strand used in Example 3.4, where $\rho = 8$ ft. In Example 3.4, the moment $m_s = 24.00$ in. lb, and, hence, Eq. (3.84) yields

$${}_m\sigma_2 = \frac{4 \times 24.00}{\pi \times 0.101)^3} = 29{,}660 \text{ psi.}$$

The maximum bending stress in the center wire is given by Eq. (3.85), where

$${}_m\sigma_1 = \frac{28{,}500{,}000 \times 0.103}{8 \times 12} = 30{,}580 \text{ psi.}$$

The maximum shearing stress, as determined by Eq. (3.81), yields

$${}_H\sigma_2 = \frac{2 \times 24.00 \times 0.13034}{\pi(0.101)^3} = 1{,}930 \text{ psi.}$$

3.7 Stress Determination of a Strand Passing over a Sheave

Since a strand wrapped around a sheave is generally restrained against rotation ($\tau_s = 0$), an axial twisting moment is developed in the strand along with the tension and torsion. Figure 3.6 shows an equilibrium configuration of a

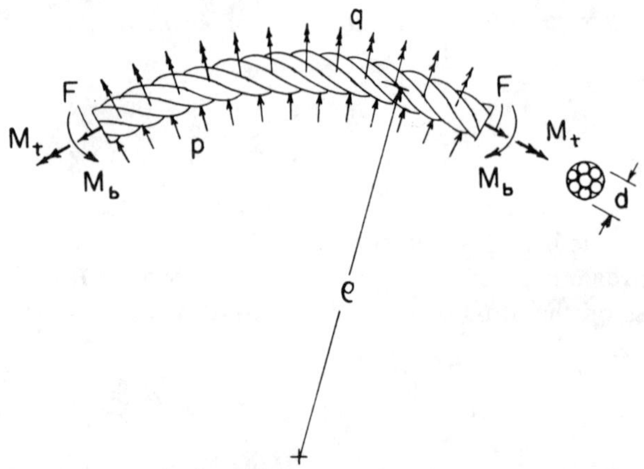

FIGURE 3.6. Simple strand wrapped around a sheave.

strand where the loads acting on any cross section are F, M_t, and M_b, which are the axial load, the axial twisting moment, and the bending moment, respectively. The loads p and q are the line load per unit length and the twisting couple per unit length, respectively, applied to the strand by the sheave in order to maintain equilibrium of the strand. Thus, the equations of equilibrium of the strand yield

$$F = p\rho \tag{3.86}$$

and

$$M_t = q\rho. \tag{3.87}$$

It should be noted that the line load per unit length acting on the strand is along the centerline of the strand.

It will now be assumed that the stresses due to the bending and axial loads are additive [9]. For example, if a thin elastic rod is loaded axially and bent over a sheave, the stresses are determined by adding the axial tension stresses to the stresses determined by pure bending. This assumption neglects the effects of friction and also the effects of the radial stresses. It should be noted that there is a transition region in the strand between the straight portion and the portion with the radius of curvature ρ. In this region, the radial loads are not as large as those in the strand on the sheave. This tends to reduce the effects of friction by possibly allowing one wire to move relative to another.

In the case of a simple strand bent over a sheave, the stresses, as mentioned previously, will be obtained by a superposition of the stresses determined in Sections 3.3 and 3.6. An illustrative example is worked out below.

Example 3.6. Consider the simple straight strand used in Example 3.1, where $R_1 = 0.103$ in., $R_2 = 0.101$ in, $p_2 = 9.75$ in., $E = 28{,}500{,}000$ psi, and $\nu = 0.25$.

Let $F = 10{,}000$ lb, $\tau_s = 0$, and $\rho = 8$ ft. From the results of Example 3.1, where for $F = 18{,}805$ lb and $\tau_s = 0$, the moment $M_t = 406$ in. lb, the moment in this case is

$$M_t = 406 \times \frac{10{,}000}{18{,}805} = 216 \text{ in. lb},$$

since, as mentioned previously, the problem is a linear one. Equations (3.86) and (3.87) result in

$$p = \frac{10{,}000}{8 \times 12} = 104.2 \text{ lb/in.}$$

and

$$q = \frac{216}{8 \times 12} = 2.25 \text{ in. lb/in.}$$

The stresses in an outer wire will be determined at the points *a*, *b*, *c*, and *d* on a cross section for various values of *s*, the arc length along the wire. Figure 3.7 shows the points *a*, *b*, *c*, and *d* of an outer wire in the unloaded configuration. These points are on the boundary with point *a* in contact with the center wire. Point *a* is in the positive direction of the *A*-axis, whereas point *b* is in the positive direction of the *B*-axis. Point *c* is opposite to point *a*, whereas point *d* is opposite to point *b*. The stresses will be determined at these points for $s = 0$, $s = \pi r_s/(2\cos\alpha_2)$, $s = \pi r_2/\cos\alpha_2$, and $s = 3\pi r_2/(2\cos\alpha_2)$. Notice that the above values of *s* correspond to $\theta = 0$, $\theta = \pi/2$, $\theta = \pi$, and $\theta = 3\pi/2$, respectively.

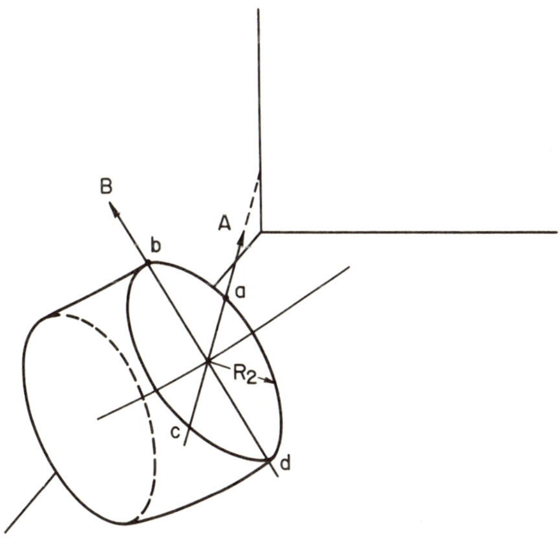

FIGURE 3.7. Cross section of an outer wire.

For the axial case and for any value of s the stresses are at

point a, $\sigma = (83{,}700 + 1{,}580) \times \dfrac{10{,}000}{18{,}805} = 45{,}350$ psi;

point b, $\sigma = 83{,}700 \times \dfrac{10{,}000}{18{,}805} = 44{,}510$ psi;

point c, $\sigma = (83{,}700 - 1{,}580) \times \dfrac{10{,}000}{18{,}805} = 43{,}670$ psi;

and at

point d, $\sigma = 83{,}700 \times \dfrac{10{,}000}{18{,}805} = 44{,}510$ psi.

For pure bending and for $s = 0$ at

point a, $\sigma = 0$ psi;

point b, $\sigma = -\dfrac{4 \times 24}{\pi \times (0.101)^3} = -29{,}660$ psi;

point c, $\sigma = 0$ psi;

and at

point d, $\sigma = 29{,}660$ psi.

For pure bending and for $s = \pi r_2/(2 \cos \alpha_2)$ at

point a, $\sigma = -\dfrac{4 \times 24 \times 0.99147}{\pi \times (0.101)^3} = -29{,}410$ psi;

point b, $\sigma = 0$ psi;

point c, $\sigma = 29{,}410$ psi;

and at

point d, $\sigma = 0$ psi.

For pure bending and for $s = \pi r_2/\cos \alpha_2$ at

point a, $\sigma = 0$ psi;

point b, $\sigma = 29{,}660$ psi;

point c, $\sigma = 0$ psi;

and at

point d, $\sigma = -29{,}660$ psi.

For pure bending and for $s = 3\pi r_2/(2 \cos \alpha_2)$ at

point a, $\sigma = 29{,}410$ psi;

TABLE 3.1. Total normal stress in an outer wire ($F = 10{,}000$ lb, $\tau_s = 0$, $\rho = 8$ ft, $R_1 = 0.103$ in., $R_2 = 0.101$ in., and $p_2 = 9.75$ in.)

s	θ	a	b	c	d
0	0	45,350	14,850	43,670	74,170
2.46	$\pi/2$	15,940	44,510	73,080	45,510
4.92	π	45,350	74,170	43,670	14,850
7.38	$3\pi/2$	74,760	44,510	14,260	44,510

point b, $\sigma = 0$ psi;

point c, $\sigma = -29{,}410$ psi;

and at

point d, $\sigma = 0$ psi.

Table 3.1 shows the total normal stress in an outer wire at the points a, b, c, and d for the indicated values of s or θ.

The maximum normal stress in the center wire is

$$\sigma_1 = 85{,}500 \times \frac{10{,}000}{18{,}805} + \frac{28{,}500{,}000 \times 0.103}{8 \times 12} = 76{,}045 \text{ psi},$$

which again is greater than the maximum tensile stress in an outer wire. The maximum shearing stress in an outer wire is

$$_H\sigma_2 = 2090 \times \frac{10{,}000}{18{,}805} + 1{,}930 = 3{,}040 \text{ psi}.$$

3.8 Multilayered Strands

The analysis presented in the previous sections can be extended to multilayered strands. Consider the addition of another layer (layer 3) of wires to the simple strand shown in Figure 3.1, where the lay of the wires in layer 3 is opposite to the lay of the wires in layer 2 ($\alpha_3 > 90°$). This is done to reduce the axial twisting moment in the strand.

An extension of the previous analysis results in the following equations [11]:

$$r_3 = R_1 + 2R_2 + R_3, \tag{3.88}$$

$$\frac{r_3}{\bar{r}_3} = 1 + \nu\frac{(R_1\xi_1 + 2R_2\xi_2 + R_3\xi_3)}{r_3}, \tag{3.89}$$

$$\xi_1 = \xi_3 + \frac{\Delta\alpha_3}{\tan\alpha_3}, \tag{3.90}$$

$$\beta_3 = r_3\tau_s = \frac{\xi_3}{\tan\alpha_3} - \Delta\alpha_3 + \nu\frac{(R_1\xi_1 + 2R_2\xi_2 + R_3\xi_3)}{r_3\tan\alpha_3}, \tag{3.91}$$

$$R_3\Delta\kappa'_3 = -\frac{2\sin\alpha_3\cos\alpha_3}{r_3/R_3}\Delta\alpha_3 + \nu\frac{(R_1\xi_1 + 2R_2\xi_2 + R_3\xi_3)\cos^2\alpha_3}{r_3r_3/R_3}, \tag{3.92}$$

$$R_3\Delta\tau_3 = \frac{(1-2\sin^2\alpha_3)}{r_3/R_3}\Delta\alpha_3 + \nu\frac{(R_1\xi_1+2R_2\xi_2+R_3\xi_3)\sin\alpha_3\cos\alpha_3}{r_3r_3/R_3}, \tag{3.93}$$

$$\frac{G'_3}{ER_3^3} = \frac{\pi}{4}R_3\Delta\kappa'_3, \tag{3.94}$$

$$\frac{H_3}{ER_3^3} = \frac{\pi}{4(1+\nu)}R_3\Delta\tau_3, \tag{3.95}$$

$$\frac{N'_3}{ER_3^2} = \frac{H_3}{ER_3^3}\frac{\cos^2\alpha_3}{r_3/R_3} - \frac{G'}{ER_3^3}\frac{\sin\alpha_3\cos\alpha_3}{r_3/R_3}, \tag{3.96}$$

$$\frac{T_3}{ER_3^2} = \pi\xi_3,$$

$$\frac{X_3}{ER_3} = \frac{N'_3}{ER_3^2}\frac{\sin\alpha_3\cos\alpha_3}{r_3/R_3} - \frac{T_3}{ER_3^2}\frac{\cos^2\alpha_3}{r_3/R_3}, \tag{3.97}$$

$$\frac{F_3}{ER_3^2} = m_3\left[\frac{T_3}{ER_3^2}\sin\alpha_3 + \frac{N_3}{ER_3^2}\cos\alpha_3\right], \tag{3.98}$$

and

$$\frac{M_3}{ERa_3^3} = m_3\left[\frac{H_3}{ER_3^3}\sin\alpha_3 + \frac{G'_3}{ER_3^3}\cos\alpha_3 + \frac{T_3}{ER_3^2}\frac{r_3}{R_3}\cos\alpha_3 - \frac{N'_3}{ER_3^2}\frac{r_3}{R_3}\sin\alpha_3\right], \tag{3.99}$$

where subscript 3 indicates those previously discussed quantities pertaining to the wires in the third layer.

The total axial force acting on the strand is

$$F = F_1 + F_2 + F_3, \tag{3.100}$$

and the total axial twisting moment is

$$M_t = M_1 + M_2 + M_3. \tag{3.101}$$

Example 3.7. Let $R_1 = 0.103$ in., $R_2 = 0.101$ in., $R_3 = 0.096$ in., $p_2 = 9.75$ in., $p_3 = 9.75$ in., $\nu = 0.25$, and $E = 30{,}000{,}000$ psi. Let the second layer be right lay and let the third layer be left lay, where $m_2 = 6$ and $m_3 = 12$. Therefore, the total radius of the strand R is

$$R = 0.103 + 2\times 0.101 + 2\times 0.096 = 0.497 \text{ in.}$$

Also $\alpha_2 = 82.51°$ and $\alpha_3 = 104.49°$; the metallic area A is given by

$$A = \pi[(0.103)^2 + 6 \times (0.101)^2 + 12 \times (0.096)^2] = 0.573 \text{ in.}^2.$$

Using the above, and proceeding in a similar manner to that used in the previous examples the following result are obtained:

$$\frac{F}{AE} = 0.9274\varepsilon - 0.0967\beta$$

and

$$\frac{M}{ER^3} = -0.2208\varepsilon + 0.0758\beta.$$

If $\varepsilon = 0.003$ and $R\tau_s = \beta = 0$, the maximum axial tensile stress in each wire is $\sigma_1 = 90{,}000$ psi, $\sigma_2 = 89{,}750$ psi, and $\sigma_3 = 86{,}090$ psi. The total axial force $F = 3000 + 16{,}795 + 28{,}038 = 47{,}830$ lb, where the loads F_1, F_2, and F_3 represent 6.3%, 35.1%, and 58.6% of the total load, respectively. Since the results are linear, a load of 30,000 lb (with $\tau_s = 0$) would produce a maximum tensile stress of $\sigma_1 = 56{,}450$ psi. Again, if $\varepsilon = 0.003$ and $\tau_s = 0$, the axial M_t is given by the above equation and hence,

$$M = -30{,}000{,}000 \times (0.497)^3 \times 0.2208 \times 0.003 = -2{,}440 \text{ in. lb.}$$

The minus sign occurs, since the outside layer is left lay.

3.9 Electric Conductor Strand

Consideration is given in this section to a simple strand consisting of a steel center wire surrounded by aluminum wires. Such a strand is used to conduct electricity and is called an ACSR (*aluminum conductor steel reinforced*) electrical conductor. The previous analysis can be extended to investigate the mechanical response of such a conductor.

Consider, for example, a simple strand consisting of a center steel wire and m_2 aluminum outside wires wrapped helically around the center steel wire. Equations (3.41) and (3.42) determine the axial force F_1 and the twisting moment M_1 in the steel wire. Hence,

$$F_1 = \pi E_s R_1^2 \xi_1 \tag{3.102}$$

and

$$M_1 = \frac{\pi E_s R_1^4 \tau_s}{4(1 + \nu_s)}, \tag{3.103}$$

where E_s is the modulus of elasticity for steel, R_1 is the radius of the center steel wire, ξ_1 is the axial strain of the conductor, τ_s is the angle of twist per unit length of the conductor and ν_s is Poisson's ratio for steel. Similarly, Eqs.

(3.34) through (3.40) become, for the outside aluminum wires,

$$\frac{G'_2}{E_a R_2^3} = \frac{\pi}{4} R_2 \Delta\kappa'_2, \tag{3.104}$$

$$\frac{H_2}{E_a R_2^3} = \frac{\pi}{4(1 + \nu_a)} R_2 \Delta\tau_2, \tag{3.105}$$

$$\frac{N'_2}{E_a R_2^2} = \frac{H_2}{E_a R_2^3} \frac{\cos^2\alpha_2}{r_2/R_2} - \frac{G'_2}{E_a R_2^3} \frac{\sin\alpha_2 \cos\alpha_2}{r_2/R_2}, \tag{3.106}$$

$$\frac{T_2}{E_a R_2^2} = \pi\xi_2, \tag{3.107}$$

$$\frac{X_2}{E_a R_2} = \frac{N'_2}{E_a R_2^2} \frac{\sin\alpha_2 \cos\alpha_2}{r_2/R_2} - \frac{T_2}{E_a R_2^2} \frac{\cos^2\alpha_2}{r_2/R_2}, \tag{3.108}$$

$$\frac{F_2}{E_a R_2^2} = m_2 \left[\frac{T_2}{E_a R_2^2} \sin\alpha_2 + \frac{N'_2}{E_a R_2^2} \cos\alpha_2 \right], \tag{3.109}$$

and

$$\frac{M_2}{E_a R_2^3} = m_2 \left[\frac{H_2}{E_a R_2^3} \sin\alpha_2 + \frac{G'_2}{E_a R_2^3} \cos\alpha_2 + \frac{T_2}{E_a R_2^2} \frac{r_2}{R_2} \cos\alpha_2 - \frac{N'_2}{E_a R_2^2} \frac{r_2}{R_2} \sin\alpha_2 \right], \tag{3.110}$$

where E_a is the modulus of elasticity for aluminum, R_2 is the aluminum wire radius, ξ_2 is the axial aluminum wire strain, and ν_a is Poisson's ratio for aluminum. Equations (3.32) and (3.33) become

$$R_2 \Delta\kappa'_2 = -\frac{2\sin\alpha_2 \cos\alpha_2}{r_2/R_2} \Delta\alpha_2 + \frac{(\nu_s R_1 \xi_1 + \nu_a R_2 \xi_2)}{r_2} \frac{\cos^2\alpha_2}{r_2/R_2} \tag{3.111}$$

and

$$R_2 \Delta\tau_2 = \frac{(1 - 2\sin^2\alpha_2)}{r_2/R_2} \Delta\alpha_2 + \frac{(\nu_s R_1 \xi_1 + \nu_a R_2 \xi_2)}{r_2} \frac{\sin\alpha_2 \cos\alpha_2}{r_2/R_2}. \tag{3.112}$$

Equation (3.27) remains the same; however, Eq. (3.31) becomes

$$r_2 \tau_s = \frac{\xi_2}{\tan\alpha_2} - \Delta\alpha_2 + \frac{(\nu_s R_1 \xi_1 + \nu_a R_2 \xi_2)}{r_2 \tan\alpha_2}. \tag{3.113}$$

The following example will illustrate the above equations.

Example 3.8. Let $m_2 = 6$, $E_s = 30 \times 10^6$ psi, $E_a = 10 \times 10^6$ psi, $R_1 = 0.067$ in., $R_2 = 0.066$ in., $\nu_s = 0.25$, $\nu_a = 0.33$, and $p_2 = 5.15$ in. Hence, $r_2 = 0.067 + 0.066 = 0.133$ in. and $\alpha_2 = 80.783°$. As before, let $\tau_s = 0$ and $\varepsilon = 0.003$. Equations (3.27) and (3.113) become

$$0.003 = \xi_2 + \frac{\Delta\alpha_2}{6.16277}$$

and

$$0 = \frac{\xi_2}{6.16277} - \Delta\alpha_2 + \frac{(0.25 \times 0.067 \times 0.003 + 0.33 \times 0.066 \times \xi_2)}{0.133 \times 6.16277},$$

which have the solution

$$\xi_2 = 0.002901 \quad \text{and} \quad \Delta\alpha_2 = 0.00060915.$$

Equations (3.111) and (3.112) yield

$$R_2 \Delta\kappa'_2 = -0.000084726 \quad \text{and} \quad R_2 \Delta\tau_2 = -0.00021986.$$

Proceeding as in the previous examples, the result is

$$F = F_1 + F_2 = 1269 + 2351 = 3620 \text{ lb}.$$

It is interesting to note that the steel wire and the aluminum wires carry 35.1% and 64.9%, respectively, of the total axial load.

Proceeding again, as in the previous examples, results in

$$F = 1.207 \times 10^6 \varepsilon + 1.694 \times 10^4 \tau_s$$

and

$$M = 1.608 \times 10^4 \varepsilon + 5.551 \times 10^2 \tau_s.$$

Again expressions can be written down for the stresses.

3.10 Contact Stresses

The line of contact between the outside wire and the center wire of a simple strand is a helix of radius R_1. An estimate of the contact stresses can be made if the contact force per unit length is known. It should be noted that the resultant force per unit length X_2 is calculated per unit length along the centerline of an outside wire and hence should be calculated per unit length of the contact line. This results in the approximate equation

$$X_c \sqrt{p_2^2 + (2\pi R_1)^2} = -X_2 \sqrt{p_2^2 + [2\pi(R_1 + R_2)]^2}, \qquad (3.114)$$

where X_c is the contact force per unit length acting along the line of contact.

Figure 3.8 shows a point along the line of contact between the two wires where the cross section of the outside wire is shown circular. In this figure, the cross section of the center wire is shown elliptical and hence the radius of curvature of the surface of the center wire, at the point of contact, is ρ_1, where

$$\rho_1 = \frac{R_1}{\sin^2 \alpha_2}. \qquad (3.115)$$

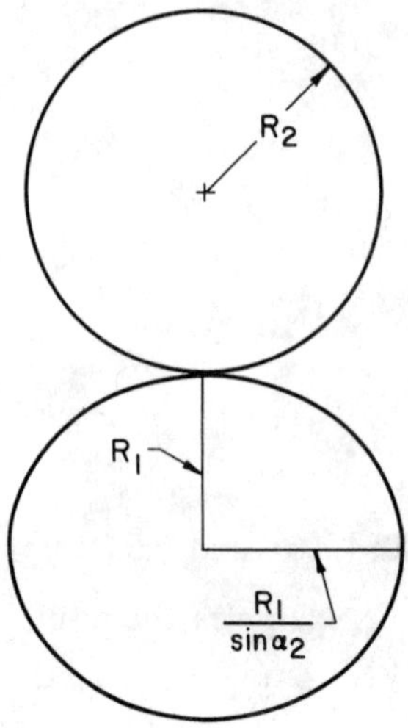

FIGURE 3.8. Contact between center and outer wire.

It will now be assumed that the contact stress is the same as that determined by two cylindrical bodies in line contact [6]. Hence, the maximum contact stress σ_c is given by the equation

$$\sigma_c = -\frac{b}{\Delta}, \tag{3.116}$$

where

$$\Delta = \frac{4(1 - v^2)}{\left[\frac{1}{\rho_1} + \frac{1}{R_2}\right]E} \tag{3.117}$$

and

$$b = \sqrt{\frac{2X_c\Delta}{\pi}}. \tag{3.118}$$

The use of the above formulae will be illustrated by the following example.

Example 3.9. Consider the simple strand used in Example 3.1, where $R_1 =$ 0.103 in., $R_2 = 0.101$ in., $\alpha_2 = 82.5106°$, $E = 28{,}500{,}000$ psi, and $v = 0.25$. In Example 3.1, an axial load of 18,805 lb with $\tau_s = 0$ produces a contact line load of

$$X_2 = -0.000077477 \times 28{,}500{,}000 \times 0.101 = -223.0 \text{ lb/in.}$$

Equation (3.114) yields

$$X_c = 223\frac{\sqrt{(9.75)^2 + (2\pi \times 0.204)^2}}{\sqrt{(9.75)^2 + (2\pi \times 0.103)^2}} = 225 \text{ lb/in.},$$

whereas Eq. (3.115) results in

$$\rho_1 = \frac{0.103}{0.98301} = 0.1048 \text{ in.}$$

The values of Δ and b are

$$\Delta = \frac{4[-(0.25)^2]}{\left(\dfrac{1}{0.1048} + \dfrac{1}{0.101}\right) \times 28{,}500{,}000} = 6.7674 \times 10^{-9} \text{ in.}^3/\text{lb}$$

and

$$b = \sqrt{\frac{2 \times 225 \times 6.7674 \times 10^{-9}}{\pi}} = 9.8456 \times 10^{-4} \text{ in.}$$

The contact stress σ_c is, therefore,

$$\sigma_c = -\frac{9.8456 \times 10^{-4}}{6.7674 \times 10^{-9}} = -145{,}900 \text{ psi.}$$

It should be noted that the contact stress is not linearly related to the axial load F. It varies as the square root of the axial load. An axial load of 10,000 lb would produce a contact stress of

$$\sigma_c = -\sqrt{\frac{10{,}000}{18{,}805}} \times 145{,}900 = -106{,}390 \text{ psi.}$$

The above stresses were calculated based on the assumption that the material remains elastic.

3.11 Outside Wires Contacting Each Other

Consider seven straight wires of circular cross parallel to each other. Let the center wire be surrounded by the six wires and let the wires have the same diameter. In this case, the outer straight wires will touch each other and the center wire simultaneously. If the outside wires are not straight but rather are wrapped helically around the center wire, the outside wires will only contact each other. This occurs since the outside wires appear elliptical in a cross section perpendicular to the center wire. Equation (3.8) can be used to demonstrate this. Contact could also be made between the outer wires if the center of the strand were made of a soft material such as a fiber.

Figure 3.9 shows the projection of the circular cross section of an outer wire on a plane perpendicular to the axis of the strand. The line of contact between adjacent wires is also a helix with radius d. An analysis of the geometry in Figure 3.9 yields [12]

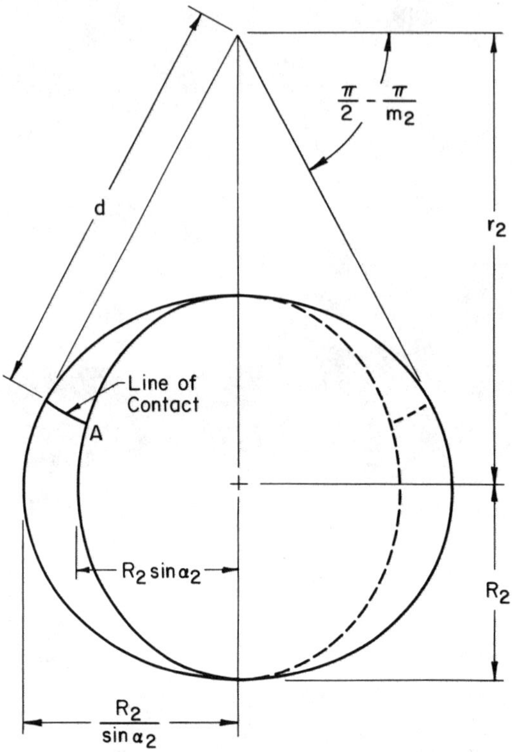

FIGURE 3.9. Line of contact between two outer wires.

$$d = \frac{R_2 \tan\left(\frac{\pi}{2} - \frac{\pi}{m_2}\right)}{\sin\alpha_2 \cos\left(\frac{\pi}{2} - \frac{\pi}{m_2}\right)\sqrt{\sin^2\alpha_2 + \tan^2\left(\frac{\pi}{2} - \frac{\pi}{m_2}\right)}}. \tag{3.119}$$

Since the projection of a circular cross section on the transverse section is elliptical, the radius d intersects the ellipse at point A, as shown in Figure 3.9.

Figure 3.10 shows a true cross-sectional view of an outer wire depicting the contact angle γ. Again an analysis of Figures 3.9 and 3.10 yields

$$\cos\gamma = \frac{1}{\cos^2\alpha_2}\left\{\sqrt{1 + \frac{\tan^2\left(\frac{\pi}{2} - \frac{\pi}{m_2}\right)}{\sin^2\alpha_2}}\right.$$
$$\left. - \sqrt{\tan^2\left(\frac{\pi}{2} - \frac{\pi}{m_2}\right)\left\{1 + \frac{1}{\tan^2\alpha_2 \cos^2\left(\frac{\pi}{2} - \frac{\pi}{m_2}\right)\left[\sin^2\alpha_2 + \tan^2\left(\frac{\pi}{2} - \frac{\pi}{m_2}\right)\right]}\right\} + \sin^4\alpha_2}\right\} \tag{3.120}$$

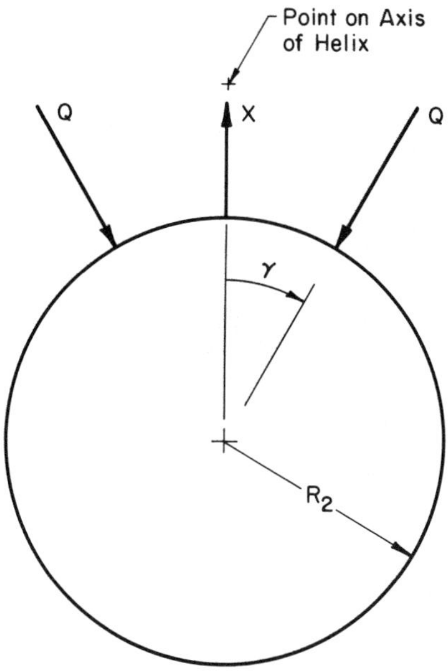

FIGURE 3.10. Contact loads on outer wire.

and, hence, the normal contact force per unit length Q is given by

$$Q = -\frac{X}{2\cos\gamma}. \tag{3.121}$$

It should be noted that, if the outside wires are touching each other, Eq. (3.1) is no longer valid. Equation (3.8) yields the correct value of r_2 when the strand is unloaded. If the strand is loaded, Eq. (38) yields the value of $\bar{r}_2$, where α becomes $\bar{\alpha}_2$ and R_2 becomes $R_2(1 - \nu\xi_2)$. Again this equation can be linearized to determine $\bar{r}_2/r_2$.

3.12 Other Types of Strand Cross Sections

In other types of strand cross sections, the preceding equations have to be slightly modified. Consider for instance the Seale strand, which consists of 19 wires as shown in Figure 3.11. Let the radius of the center wire be R_5. The two outer layers will have wire radii of R_6 and R_7 as shown in Figure 3.12.

The value of r_6 is given by the expression

$$r_6 = R_5 + R_6. \tag{3.122}$$

The value of r_7 will be calculated by assuming that the cross sections of the outside wires appear circular in Figure 3.12. From Figure 3.12 there results

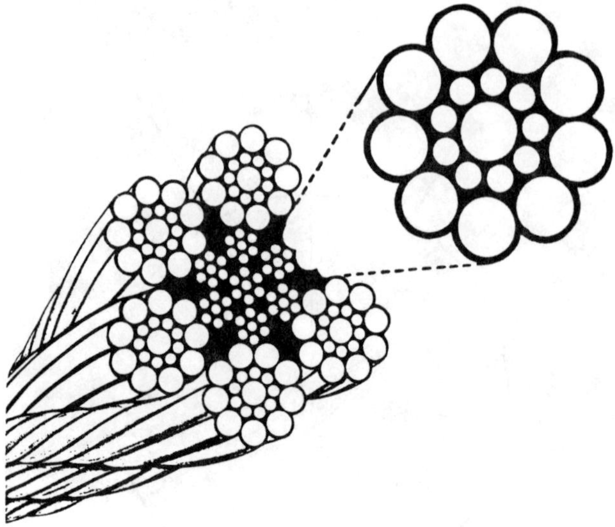

FIGURE 3.11. Cross section of a Seale wire rope.

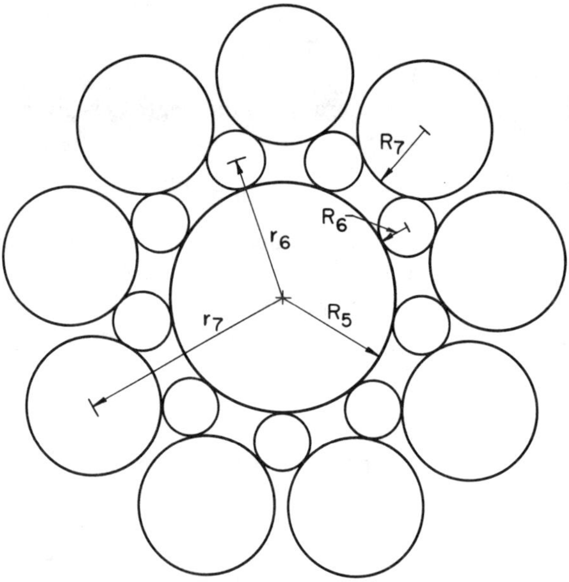

FIGURE 3.12. Cross section of a Seale strand.

$$r_7 = (R_5 + R_6)\cos 20^\circ + \sqrt{(R_6 + R_7)^2 - [(R_5 + R_6)\sin 20^\circ]^2}. \quad (3.123)$$

When the strand is loaded, $\bar{r}_7$ will change due to a Poisson's ratio effect and a change in helix angle. However, since the wires are assumed to appear circular, the value of $\bar{r}_7$ will be given by the expression

$$\begin{aligned} r^{-}{}_7 = {} & [R_5(1-v\xi_5)+R_6(1-v\xi_6)]\cos 20^\circ \\ & +\sqrt{[R_6(1-v\xi_6)+R_7(1-v\xi_7)]^2-\{[R_5(1-v\xi_5)+R_6(1-v\xi_6)]\sin 20^\circ\}^2}. \end{aligned} \quad (3.124)$$

Again $\bar{r}_7$ can be linearized with the result that

$$\begin{aligned} \bar{r}_7 = r_7 - v\Bigg\{ & (R_5\xi_5 + R_6\xi_6)\cos 20^\circ \\ & + \frac{(R_6 + R_7)(R_6\xi_6 + R_7\xi_7) - (R_5 + R_6)(R_5\xi_5 + R_6\xi_6)\sin^2 20^\circ}{\sqrt{(R_6 + R_7)^2 - [(R_5 + R_6)\sin 20^\circ]^2}}\Bigg\} \end{aligned} \quad (3.125)$$

A similar procedure can be followed with other types of cross sections.

4
Static Response of a Wire Rope

4.1 Axial Response of an Independent Wire Rope Core

The equations developed in the previous sections will now be used to generate the solution for ropes with complex cross sections. Consideration will be given, at first, to a cross section consisting of one simple, straight, seven-wire strand surrounded by six seven-wire strands, which are preformed. Figure 4.1 shows such a cross section in which the outside strands are deformed into the helical shape they assume in the unloaded rope and are this shape under no external loads (preformed strands). Such a cross section is often used as a rope core in a more complex rope and as such is sometimes called an *independent wire rope core* (IWRC). The center strand will be called strand 1, an outside strand will be denoted strand 2.

The bending stiffness of strand 2 will be approximated by a summation of the bending stiffness of each wire in the strand. Therefore, the bending stiffness, A^*_2, for strand 2, is given by the expression [see Eq. (3.78)]

$$A^*_2 = 6\frac{\pi E R_4^4}{4}\frac{2\sin\alpha_4}{(2+\nu\cos^2\alpha_4)} + \frac{\pi E R_3^4}{4}, \tag{4.1}$$

where R_3 is the center wire radius in strand 2, R_4 is an outer wire radius in strand 2, and α_4 is the helix angle of an outer wire in strand 2.

In the foregoing theory, for the solid wire strands in strand 1, the twisting moment and the axial force in the strand wires are determined from the properties of a straight solid wire with an angle of twist per unit length of $\Delta\tau_2$ and an axial straing of ξ_2, that is, $H = \pi E R_2^4 \Delta\tau_2/4(1+\nu)$ and $T = \pi E R_2^2 \xi_2$. Thus, if the cross section of a strand in a rope resembles that shown in Figure 3.1, the axial strain and the angle of twist per unit length will be used to determine the axial force and the axial twisting moment in the curved strand [13].

Let the helix angle of strand 2, shown in Figure 4.1, be α^*_2. As the rope is loaded, this helix angle assumes a new value $\bar{\alpha}^*_2$. The angle of twist per unit length for strand 2 becomes

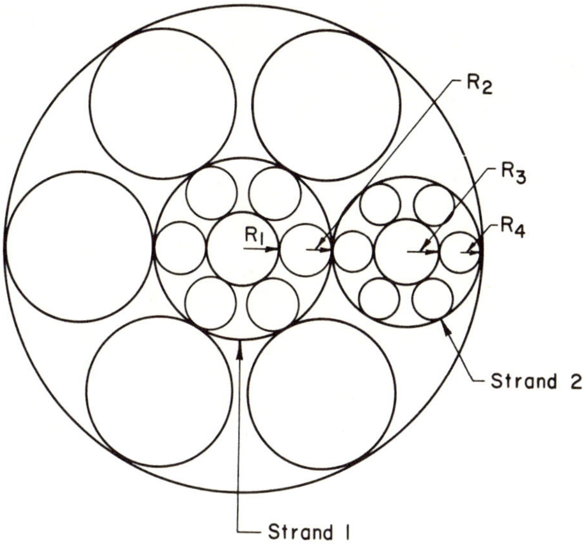

FIGURE 4.1. Independent wire rope core (IWRC).

$$\Delta\tau^*{}_2 = \frac{\sin\bar{\alpha}^*{}_2 \cos\bar{\alpha}^*{}_2}{\bar{r}^*{}_2} - \frac{\sin\alpha^*{}_2 \cos\alpha^*{}_2}{r^*{}_2}, \tag{4.2}$$

in which

$$r^*{}_2 = R_1 + 2R_2 + 2R_4 + R_3, \tag{4.3}$$

and where, due to the Poisson's ratio effect,

$$\bar{r}^*{}_2 = r^*{}_2 - \nu(R_1\xi_1 + 2R_2\xi_2 + 2R_4\xi_4 + R_3\xi_3), \tag{4.4}$$

in which ξ_1, ξ_2, ξ_3, and ξ_4 are the axial wire strains in wires 1, 2, 3, and 4, respectively. Again, Eq. (4.2) can be linearized.

Guided by the analysis of strand 1, the following equations can be written:

$$\xi_1 = \xi_3 + \frac{\Delta\alpha^*{}_2}{\tan\alpha^*{}_2}, \tag{4.5}$$

$$\xi_3 = \xi_4 + \frac{\Delta\alpha_4}{\tan\alpha_4}, \tag{4.6}$$

$$\begin{aligned} r^*{}_2\tau &= \frac{r^*{}_2}{\bar{r}^*{}_2}\left(\frac{(1+\xi_3)}{\tan\alpha^*{}_2} - \Delta\alpha^*{}_2\right) - \frac{1}{\tan\alpha^*{}_2} \\ &= \frac{\xi_3}{\tan\alpha^*{}_2} - \Delta\alpha^*{}_2 + \frac{\nu}{r^*{}_2}\frac{(R_1\xi_1 + 2R_2\xi_2 + 2R_4\xi_4 + R_3\xi_3)}{\tan\alpha^*{}_2}, \end{aligned} \tag{4.7}$$

and

$$(R_3 + R_4)\Delta\tau^*_2 = \frac{\xi_4}{\tan\alpha_4} - \Delta\alpha_4 + \nu\frac{(R_3\xi_3 + R_4\xi_4)}{(R_3 + R_4)\tan\alpha_4}$$
$$= \frac{(R_3 + R_4)}{r^*_2}\left[(1 - 2\sin^2\alpha^*_2)\Delta\alpha^*_2 + \nu\frac{(R_1\xi_1 + 2R_2\xi_2 + 2R_4\xi_4 + R_3\xi_3)}{r^*_2}\sin\alpha^*_2\cos\alpha^*_2\right], \tag{4.8}$$

where τ is the twist per unit length of the rope and $\Delta\tau^*_2$ is the angle of twist per unit length of strand 2. Also for strand 1,

$$\xi_1 = \xi_2 + \frac{\Delta\alpha_2}{\tan\alpha_2} \tag{4.9}$$

and

$$(R_1 + R_2)\tau = \frac{\xi_2}{\tan\alpha_2} - \Delta\alpha_2 + \nu\frac{(R_1\xi_1 + R_2\xi_2)}{(R_1 + R_2)\tan\alpha_2}. \tag{4.10}$$

Equation (3.53) yields the rotational strain. Therefore, for the rope,

$$\beta = R\tau, \tag{4.11}$$

where

$$R = R_1 + 2R_2 + 2R_3 + 4R_4. \tag{4.12}$$

The following procedure can now be used to determine the axial response of the rope shown in Figure 4.1. Choose values of ε and β, which are the axial and rotational strain of the rope. Since $\varepsilon = \xi_1$ and $\beta = R\tau$, Eqs. (4.9) and (4.10) can be used to solve for ξ_2 and $\Delta\alpha_2$. Equations (4.5) through (4.8) can now be solved for ξ_3, ξ_4, $\Delta\alpha^*_2$, and $\Delta\alpha_4$. Once ξ_3 and $\Delta\tau^*_2$ [Eq. (4.8)] are known, the total axial force T^*_2 and the total axial twisting moment H^*_2 in strand 2 can be determined by utilizing the equations for a strand in Chapter 3. The bending moment $G_2'^*$ in strand 2 is given by the expression

$$G_2'^* = A^*_2\Delta\kappa^*_2 = A^*_2\left(\frac{\cos^2\bar{\alpha}^*_2}{r^*_2} - \frac{\cos^2\alpha^*_2}{r^*_2}\right), \tag{4.13}$$

which again can be linearized. Equation (3.36) yields the value of $N'_2{}^*$, where

$$N'_2{}^* = H^*_2\frac{\cos^2\alpha^*_2}{r^*_2} - G'_2{}^*\frac{\sin\alpha^*_2\cos\alpha^*_2}{r^*_2} \tag{4.14}$$

and Eqs. (3.39) and (3.40) determine, in the axial direction of the rope, the axial force and axial twisting moment of the number 2 strands. Hence,

$$F^*_2 = 6(T^*_2\sin\alpha^*_2 + N'_2{}^*\cos\alpha^*_2) \tag{4.15}$$

and

$$M^*_{t2} = 6(H^*_2 \sin\alpha^*_2 + G_2'^* \cos\alpha^*_2 + T^*_2 r^*_2 \cos\alpha^*_2 - N'_2{}^* r^*_2 \sin\alpha^*_2). \tag{4.16}$$

The total axial force and axial twisting moment acting on the rope would, of course, be

$$F = F^*_1 + F^*_2 \tag{4.17}$$

and

$$M_t = M^*_{t1} + M^*_{t2}. \tag{4.18}$$

Since the linearized form of Eq. (4.13) will be needed, it is listed below. Hence,

$$\begin{aligned} G'_2{}^* &= A^*_2 \Delta\kappa^*_2 = A^*_2\left(\frac{\cos^2\bar{\alpha}^*_2}{\bar{r}^*_2} - \frac{\cos^2\alpha^*_2}{r^*_2}\right) \\ &= A^*_2\left[\nu\frac{\cos^2\alpha^*_2}{(r^*_2)^2}(R_1\xi_1 + 2R_2\xi_2 + 2R_4\xi_4 + R_3\xi_3)\right. \\ &\quad \left. - \frac{2\sin\alpha^*_2\cos\alpha^*_2}{r^*_2}\Delta\alpha^*_2\right]. \end{aligned} \tag{4.19}$$

The preceding equations will now be used in an example to determine the axial response of an independent wire rope core as shown in Figure 4.1. It should be noted that the above equations could be progammed on a computer to facilitate their solution. However, it is left that a numerical example should be worked out first, to trace the steps involved in the solution.

Example 4.1. Let $R_1 = 0.0315$ in., $R_2 = 0.028925$ in., $R_3 = 0.027725$ in., $R_4 = 0.025815$ in., $E = 30 \times 10^6$ psi, $\nu = 0.29$, $p_2 = 1.30$ in., $p_4 = 2.14$ in., and $p^*_2 = 3.05$ in.

Strand 1 will be considered first. A check will be made to see if the outside wires in strand 1 are touching each other. If it is assumed that the wires are not touching each other, then Eq. (3.1) yields

$$r_2 = 0.03155 + 0.028925 = 0.060475 \text{ in.}$$

Equation (3.10) yields

$$\tan\alpha_2 = \frac{1.30}{2\pi \times 0.060475} = 3.421272,$$

and hence $\alpha_2 = 73.70694°$. With this value of α_2, Eq. (3.8) gives

$$r = 0.028925\sqrt{1 + \frac{\tan^2\left(\dfrac{\pi}{2} - \dfrac{\pi}{6}\right)}{0.9212915}} = 0.05967 \text{ in.},$$

and since 0.05967 in. $<$ 0.060475 in., Eq. (3.9) indicates that the wires are not touching each other. Therefore, $\alpha_2 = 73.70694°$ and $r_2 = 0.060475$ in.

Equations (3.50) and (3.51) can now be determined for strand 1 and the procedure used is the same as that used in Chapter 3. Hence, for strand 1, the following results,

$$\frac{F}{AE} = 0.8864\varepsilon + 0.1436\beta \tag{a}$$

and

$$\frac{M_t}{ER^3} = 0.3189\varepsilon + 0.1183\beta, \tag{b}$$

are obtained, whereas for strand 2,

$$\frac{F}{AE} = 0.9642\varepsilon + 0.0853\beta \tag{c}$$

and

$$\frac{M_t}{ER^3} = 0.1928\varepsilon + 0.0721\beta. \tag{d}$$

The bending stiffness of strand 2 is given by Eq. (4.1), and therefore

$$A^*{}_2 = \frac{6\pi \times 30 \times 10^6 \times (0.025815)^4 \times 0.98787}{2(2 + 0.29 \times 0.024115)} + \frac{\pi \times 30 \times 10^6 (0.027725)^4}{4}$$

$$= 75.728 \text{ lb in.}^2.$$

It should be mentioned that a check of strand 2 shows that the wires in strand 2 do not touch each other. Also a check of strand 2 shows that strands 2 do not touch each other.

Now let $\varepsilon = 0.0015$ and $\tau = 0$ for the rope. Equations (4.9) and (4.10) become

$$0.0015 = \xi_2 + \frac{\Delta\alpha_2}{3.42127}$$

and

$$0 = \frac{\xi_2}{3.42127} - \Delta\alpha_2 + \frac{0.29(0.03155 \times 0.0015 + 0.028925 \times \xi_2)}{0.060475 \times 3.42127},$$

which have the solution $\xi_2 = 0.001349$ and $\Delta\alpha_2 = 0.0005154$. Equations (4.5) through (4.8) become

$$0.0015 = \xi_3 + \frac{\Delta\alpha^*{}_2}{2.87649}$$

where

$$\xi_3 = \xi_4 + \frac{\Delta\alpha_4}{6.36144},$$

$$0 = \frac{\xi_3}{2.87649} - \Delta\alpha^*{}_2 + \frac{0.29(0.000047325 + 0.000078059 + 0.05163\xi_4 + 0.027725\xi_3)}{0.16876 \times 2.87649},$$

and

$$\frac{\xi_4}{6.36144} - \Delta\alpha_4 + 0.29\frac{(0.027725\xi_3 + 0.025815\xi_4)}{0.05354 \times 6.36149}$$
$$= \frac{0.05354}{0.16876}\left[(1 - 2 \times 0.892174)\Delta\alpha^*{}_2 + \frac{0.29}{0.16876}(0.000047325 + 0.000078059 + 0.05163\xi_4 + 0.027725\xi_3)0.3101604\right],$$

which have the solution $\xi_3 = 0.0012965$, $\xi_4 = 0.0012399$, $\Delta\alpha_4 = 0.00036032$, and $\Delta\alpha^*{}_4 = 0.00058535$. Equation (4.8) can be solved for $\Delta\tau^*{}_2$ and yields

$$\Delta\tau^*{}_2 = \frac{1}{0.05354}\left[\frac{0.0012399}{6.36144} - 0.00036032 + \frac{0.29(0.000035945 + 0.00003201)}{0.05354 \times 6.36144}\right]$$
$$= -0.002009 \text{ in.}^{-1},$$

and, hence, β for strand 2 is $-0.079355 \times 0.002009 = -0.000159413$.

The metallic,cross-sectional areas of strand 1 and strand 2 are 0.0188977 in.^{2} and 0.01497648 in.^{2}, respectively. The axial force and axial twisting moment in strand 1 can be computed from Eqs. (a) and (b), above. Hence,

$$F^*{}_1 = 30 \times 10^6 \times 0.0188977 \times 0.8861 \times 0.0015 = 753.83 \text{ lb}$$

and

$$M^*{}_{t1} = 30 \times 10^6 \times (0.0894)^3 \times 0.3189 \times 0.0015 = 10.25 \text{ in. lb}.$$

Also for strand 2, Eqs. (c) and (d), above, result in

$$T^*{}_2 = 30 \times 10^6 \times 0.01498765(0.9642 \times 0.0012965 - 0.0853 \times 0.000159413)$$
$$= 555.53 \text{ lb}$$

and

$$H^*{}_2 = 30 \times 10^6 \times (0.079355)^3(0.1928 \times 0.0012965 - 0.0721 \times 0.000159413)$$
$$= 3.575 \text{ in. lb}.$$

Equation (4.19) can now be used to compute $G'_2{}^*$. Therefore,

$$G'_2{}^* = 75.728\left[\frac{0.29(0.32837)^2}{(0.16876)^2}(0.03155 \times 0.0015 + 2 \times 0.028925 \times 0.001349 + 2 \times 0.025815 \times 0.0012399 + 0.027725 \times 0.0012965) - \frac{2 \times 0.94455 \times 0.32837 \times 0.00058535}{0.16876}\right]$$

$$= -0.14420 \text{ in. lb.}$$

$N'_2{}^*$ is given by Eq. (4.14), and hence,

$$N'_2{}^* = \frac{3.575(0.32837)^2}{0.16876} + \frac{0.1442 \times 0.94455 \times 0.32837}{0.16876} = 2.550 \text{ lb.}$$

Since the values of T^*_2, $N'_2{}^*$, $G'_2{}^*$, and H^*_2 are known, Eqs. (4.15) and (4.16) can be used to determine F^*_2 and M^*_{t2}. Hence,

$$F^*_2 = 6[555.53 \times 0.94455 + 2.550 \times 0.32837] = 3153.4 \text{ lb}$$

and

$$M^*_{t2} = 6[3.575 \times 0.94455 - 0.14420 \times 0.32837 + 555.53 \times 0.16876 \times 0.32837 - 2.550 \times 0.16876 \times 0.94455]$$

$$= 202.2 \text{ in. lb.}$$

Equations (4.17) and (4.18) yield the total axial force and axial twisting moment acting on the rope. Therefore,

$$F = 753.8 + 3153.4 = 3907.2 \text{ lb}$$

and

$$M_t = 10.3 + 202.2 = 212.5 \text{ in. lb.}$$

The total metallic cross-sectional area of the rope [Eq. (3.52)] is

$$A = 0.10876 \text{ in.}^2,$$

whereas the radius of the rope is [Eq. (4.12)]

$$R = 0.2481 \text{ in.}$$

Hence, Eqs. (3.50) and (3.51) become (since $\beta = 0$)

$$\frac{3907.2}{0.10876 \times 30 \times 10^6} = C_1 \times 0.0015 + C_2 \times 0$$

and

$$\frac{212.5}{30 \times 10^6 \times (0.2481)^3} = C_3 \times 0.0015 + C_4 \times 0.$$

Hence,

$$C_1 = 0.798 \quad \text{and} \quad C_3 = 0.309.$$

A similar procedure, where $\varepsilon = 0$ and β takes on a given value, yields the values of C_2 and C_4. Therefore, the following results:

$$\frac{F}{AE} = 0.798\varepsilon + 0.180\beta$$

and

$$\frac{M_t}{ER^3} = 0.309\varepsilon + 0.084\beta.$$

4.2 Axial Response of a Wire Rope

It is obvious how the preceding equations could be extended to a wire rope with more complex cross sections. Figure 4.2 shows such a cross section. It is called a 6 × 19 Seale rope with an IWRC.

It is instructive at this point to discuss the results of applying the previous theory to the Seale rope shown in Figure 4.2. Let $R_1 = 0.03155$ in., $R_2 = 0.028925$ in., $R_3 = 0.027725$ in., $R_4 = 0.025815$ in., $R_5 = 0.05731$ in., $R_6 = 0.02805$ in., $R_7 = 0.049928$ in., $\alpha_2 = 73.7069°$, $\alpha_4 = 81.0664°$, $\alpha_6 = 102.2670°$, $\alpha_7 = 111.2312°$, $\alpha^*{}_2 = 70.8302°$, and $\alpha^*{}_3 = 70.2389°$. The following results from an applicationof the preceding theory [14]:

$$\left.\begin{aligned} \frac{F}{AE} &= 0.8864\varepsilon + 0.1436\beta \\ \frac{M}{ER^3} &= 0.3189\varepsilon + 0.1182\beta \end{aligned}\right\} \quad \text{strand 1} \qquad \text{(a)}$$

$$\left.\begin{aligned} \frac{F}{AE} &= 0.9642\varepsilon + 0.0853\beta \\ \frac{M}{ER^3} &= 0.1928\varepsilon + 0.0721\beta \end{aligned}\right\} \quad \text{strand 2} \qquad \text{(b)}$$

$$\left.\begin{aligned} \frac{F}{AE} &= 0.8295\varepsilon - 0.1765\beta \\ \frac{M}{ER^3} &= -0.4240\varepsilon + 0.1480\beta \end{aligned}\right\} \quad \text{strand 3} \qquad \text{(c)}$$

Notice that the value of C_1 for strand 2 is 0.9642 and the value of C_1 for strand 3 is 0.8295. This is mainly due to the fact that α_4 is closer to 90° than is α_6 and α_7 and, hence, strand 2 is stiffer than strand 3. As noted before when strands

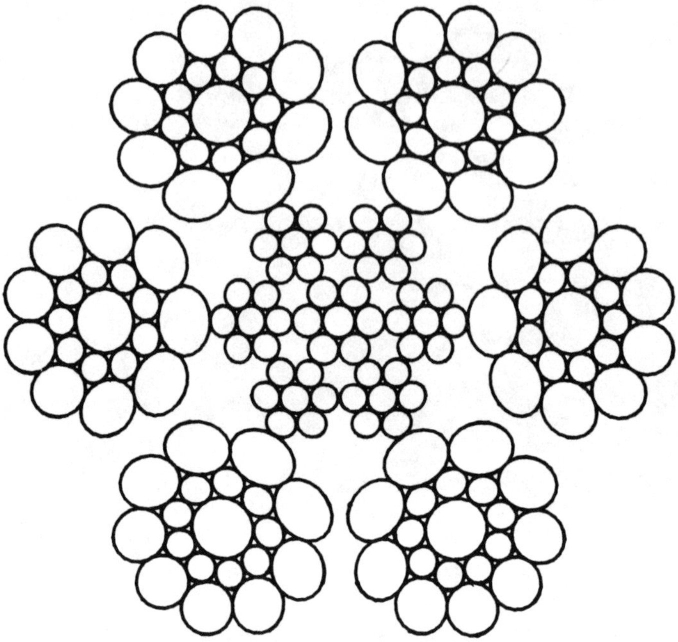

FIGURE 4.2. 6 × 19 Seale wire rope (IWRC).

1 and 2 are placed together to form the independent wire rope core, the following results:

$$\frac{F}{AE} = 0.7984\varepsilon + 0.1799\beta \tag{d}$$

and

$$\frac{M}{ER^3} = 0.3092\varepsilon + 0.0840\beta. \tag{e}$$

The value of C_1 has now dropped to 0.7984.

The minus signs occur in Eq. (c), above, since the outer wires in strand 3 are left lay. When strands 1, 2, and 3 are placed together to form the 6 × 19 Seale IWRC rope, there results

$$\frac{F}{AE} = 0.7020\varepsilon + 0.1232\beta \tag{f}$$

and

$$\frac{M}{ER^3} = 0.2060\varepsilon + 0.0403\beta. \tag{g}$$

Notice now that C_1 has dropped to 0.702. This result will be compared with a test conducted on a 1.306 in. diameter, 6 × 19 Seale IWRC wire rope in Chapter 6.

It is interesting to compare the various wire strains ($\xi_1, \ldots, \xi_7$) for the Seale rope. The values are (for $\varepsilon = 0.0015$ and $\beta = 0$) $\xi_1 = 0.0015$, $\xi_2 = 0.00135$, $\xi_3 = 0.00130$, $\xi_4 = 0.0124$, $\xi_5 = 0.00129$, $\xi_6 = 0.00123$, and $\xi_7 = 0.00111$. These results again show that, under no rotation, the maximum axial wire strain occurs in the center wire. The individual strand lays are responsible for 3.78, 13.74, and 82.98% of the total axial load. Therefore, the IWRC carries approximately 17% of the total rope load [14].

4.3 Stresses in a Wire Rope Due to Axial Loading

Consideration will be given at first to the IWRC. In the case of the center strand (strand 1), a knowledge of the axial and rotational strain, as determined by the previous section, yields the axial loads F^*_1 and M^*_{t1}. Also, the axial and rotational strain for strand 1 determine the wire stresses in strand 1, as shown in Chapter 3, Section 3.3. Strand 2 has, from the previous section, known axial loads T^*_2 and H^*_2 and a known bending moment $G'_2{}^*$. Again the stresses in strand 2 can be determined by the methods employed in Chapter 3, Section 3.7.

It is worth noting that a simple strand behaves like a center wire surrounded by constrained helical springs (outer wires). In bending and in tension it is generally the center wire that receives the largest stresses (provided the twisting moment is not excessive). Hence, attention will be focused on the center wire of strand 1 and the center wire of strand 2.

For strand 1, with $\beta = 0$ (no rotation of the rope), the maximum axial stress is simply

$$\sigma_1 = E\varepsilon. \tag{4.20}$$

For strand 2, the maximum axial stress is due to the axial strain ξ_3 and the strain due to the change in curvature $\Delta\kappa^*_2$. Hence, the maximum normal stress in the center wire of strand 2 is

$$\sigma_2 = E\xi_3 + ER_3\Delta\kappa^*_2, \tag{4.21}$$

where $\Delta\kappa^*_2$ is determined from Eq. (4.19). These stresses will be determined for the example considered below.

Example 4.2. Consider the IWRC used in Example 4.1. Let the axial load be 3,907 lb and let $\beta = 0$. The axial strain ε is 0.0015, and hence, the maximum axial stress in the center wire is [Eq. (4.20)]

$$\sigma_1 = 30 \times 10^6 \times 0.0015 = 45{,}000 \text{ psi}.$$

Since $G'_2{}^* = A^*_2\Delta\kappa^*_2$, the change in curvature is

$$\Delta\kappa^*{}_2 = \frac{0.14420}{75.728} = 0.001904 \text{ in.}^{-1},$$

and, hence, the maximum normal stress in the center wire of strand 2 is [Eq. (4.21)]

$$\sigma_2 = 30 \times 10^6 \times 0.001297 + 30 \times 10^6 \times 0.027725 \times 0.001904$$
$$= 38{,}910 + 1584 = 40{,}494 \text{ psi},$$

which again is less than the stress in the center wire of strand 1.

It is interesting to compare the maximum normal stress in the center wire of strand 1 with the stress computed by taking the total axial load and dividing it by the metallic area. This yields the nominal stress

$$\sigma_{\text{nom}} = \frac{3907}{0.1088} = 35{,}910 \text{ psi},$$

which is considerably less than the 45,000 psi in the center wire.

As was mentioned previously, the stresses in the outer wires of strand 2 can be determined by treating strand 2 as a straight strand with axial loads $T^*{}_2$ and $H^*{}_2$ and bending moment $G'{}_2{}^*$. The bending moment $G'{}_2{}^*$ produces a change in curvature $\Delta\kappa^*{}_2$, and thus Section 3.6 can be used to determine the stresses due to this bending. As is shown in Section 3.6, the stresses depend upon arc length along an outer wire and on position in the cross section and this thus complicates the problem. This variation in the bending stress is not significant in comparison to the maximum stress in an outer wire, and hence, as a conservative estimate, the maximum bending stress in an outer wire, due to the change in curvature $\Delta\kappa^*{}_2$, will be added on to those produced by $T^*{}_2$ and $H^*{}_2$. This stress, due to $\Delta\kappa^*{}_2$, is given by the expression [see Eq. (3.76)]

$$\sigma = \frac{2 \sin \alpha_4}{(2 + \nu \cos^2 \alpha_4)} \Delta\kappa^*{}_2 E R_4. \tag{4.22}$$

The work above considered the stresses in an IWRC. Similar expressions can be written for a Seale rope with an IWRC. It is worth noting that the Seale strand has the largest wire radius (R_5) of all the wires in the rope and, hence, should be checked for the bending stresses.

4.4 Stresses in a Wire Rope Due to Axial Loading and Bending

The previous section dealt with stresses in ropes due to axial loads only. To obtain the stresses in a rope which, in addition to axial loads, is subjected to bending, the principle of super position will again be used, that is, the stresses due to bending will be added to those in the axial case.

In the case of a spring subjected to pure bending, Eq. (3.76) indicates that

$$m_s = EI\Delta\kappa = EI\frac{2\sin\alpha}{(2 + v\cos^2 v)}\frac{1}{\rho}, \tag{4.23}$$

where m_s is the bending moment applied to the spring, $\Delta\kappa$ is the maximum change in curvature of the wire (since m_s is the maximum bending moment on a wire cross section), E is the modulus of elasticity of the wire material, I is the moment of inertia of the wire cross section, α is the helix angle, v is Poisson's ratio, and ρ is the radius of curvature of the centerline of the spring. This means that because of the helix angle the maximum change in curvature in an outer wire is equal to the product of $2\sin\alpha/(2 + v\cos^2\alpha)$ times, the change in curvature of the centerline $1/\rho$. In the case of a rope, the wires in an outer layer have the shape of a helix on a helix. Hence, the change in curvature of such a wire will be taken as

$$\Delta\kappa = \left(\frac{2\sin\alpha}{2 + v\cos^2\alpha}\right)\left(\frac{2\sin\alpha^*}{2 + v^*\cos^2\alpha^*}\right)\frac{1}{\rho}, \tag{4.24}$$

where ρ is the radius of curvature of the centerline of the rope, α^* is the helix angle of the strand, v^* is a Poisson's ratio of a strand (which will be taken equal to v, since the term $v^*\cos^2\alpha^*$ is very small compared with 2), α is the helix angle of the wire in the strand, and v is Poisson's ratio of the wire material. It is this change in curvature that will be used to calculate the stresses due to bending of the rope.

Consider a straight thin rod of diameter d bent over a sheave of diameter D. Let the axial load on the rod be F. The maximum normal stress in the rod is

$$\sigma = \frac{4F}{\pi d^2} + E\frac{d}{D}, \tag{4.25}$$

which can be written as

$$\frac{\sigma}{\sigma_{\text{nom}}} = 1 + \frac{Ed}{\sigma_{\text{nom}}D} = 1 + \frac{1}{\dfrac{\sigma_{\text{nom}}}{E} \times \dfrac{D}{d}}, \tag{4.26}$$

where the nominal stress σ_{nom} is given by the expression

$$\sigma_{\text{nom}} = \frac{4F}{\pi d^2}. \tag{4.27}$$

Equation (4.26) shows that $\sigma/\sigma_{\text{nom}}$ can be expressed as a function of the variable $\sigma_{\text{nom}}D/Ed$. An equation similar to Eq. (4.26) can be determined for a rope with a complex cross section. There is computer program for such a result [15]. A plot of the maximum normal wire stress for each wire as a function of $\sigma_{\text{nom}}D/Ed$ is shown in Figures 4.3 and 4.4. Figure 4.3 is for a 6×19 Seale IWRC (7×7 core), whereas Figure 4.4 is for a 6×25 F filler-wire IWRC. Use

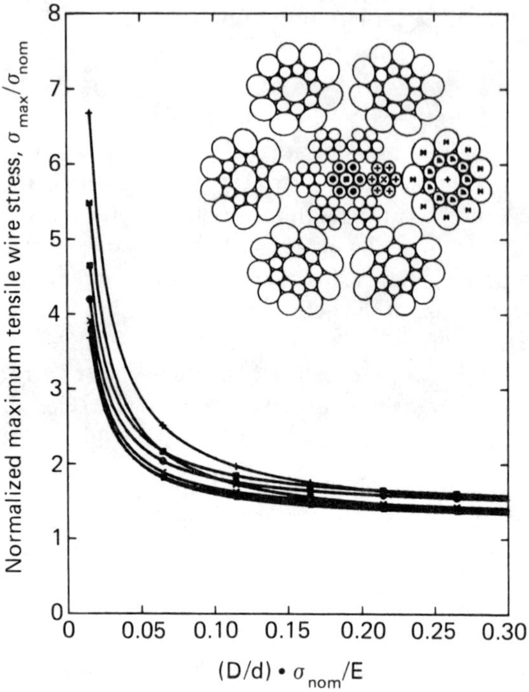

FIGURE 4.3. Maximum normal wire stresses as a function of load and curvature (6 × 19 Seale IWRC, 7 × 7 core).

of the figures will be demonstrated in the example worked out below. It should be remembered that these figures are valid for a rope that is not allowed to rotate.

Example 4.3. Consider a Seale 6 × 19 IWRC with a metallic area of 1.00 in.^{2} and an axial load of 25 tons. The nominal stress then becomes

$$\sigma_{\text{nom}} = \frac{25 \times 2000}{1} = 50{,}000 \text{ psi.}$$

Then if the D/d ratio is 30 and the rope is made of steel, the corresponding value of $\dfrac{D}{d} \times \dfrac{\sigma_{\text{nom}}}{E}$ is

$$\frac{D}{d} \times \frac{\sigma_{\text{nom}}}{E} = 0.05,$$

and, hence, from Figure 4.3, the maximum stress, which occurs in the center wire of the Seale strand, is

$$\sigma = 2.9 \times 50{,}000 = 145{,}000 \text{ psi.}$$

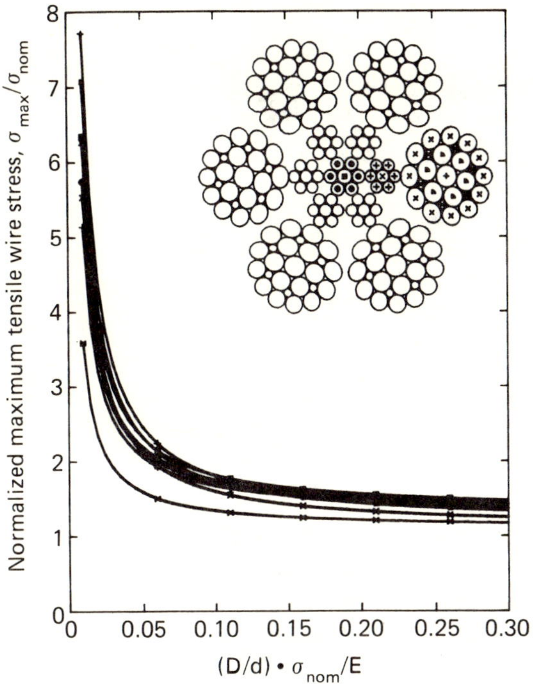

FIGURE 4.4. Maximum normal wire stresses as a function of load and curvature (6 × 25*F* filler wire IWRC).

It should be noted that in Figures 4.3 and 4.4 some of the curves cross one another. For a large radius of curvature, the center wire suffers the greatest stress under the axial load. As the radius of curvature becomes smaller, the bending stresses dominate in the wires and, hence, the larger-diameter wires will receive the largest stresses.

5
Friction in Wire Rope

5.1 Friction in an Axially Loaded Strand

Chapter 3 considered the static response of a simple straight strand with the cross section shown in Figure 3.1. It was assumed in that chapter that the wires were frictionless. It will now be shown that friction plays a very small role in the axial loading of a simple strand.

Consider an infinitesimal line element in an outer wire cross section that, if extended, passes through the centerline of the straight center wire in the unloaded configuration. This is the line element "a" discussed in Chapter 2. When the strand is axially loaded with a force and twisting moment, this line element, if extended, still passes through the centerline of the straight center strand. The shearing force N_2 in an outer wire cross section as determined by Eq. (3.17) is equal to zero. This means that all radial lines eminating from the centerline of the center straight wire and passing through the centerline of the outer wires remain straight under the axial loading. There is no relative motion between the points of contact of an outer wire with the center wire. This results in no frictional loads even if the material has a coefficient of friction.

In the case of a multilayered strand in which the lays are alternate, there will be a small effect due to friction. At the contact points, a deformation is caused by contact stresses. As the strand is loaded, there is a rotation of one wire relative to another at the contact points about a normal to the contact points, since the wires tend to straighten out under an axial load. This will be a very small effect as far as the axial response of the rope is concerned.

5.2 Frictional Effects in the Bending and Axial Loading of a Strand

Consider a simple strand that is under an axial load and is then subjected to bending. It is assumed that the axial load is sufficiently large so that contact is maintained between an outer wire and the center wire. The analysis will again involve a superposition of the stresses caused by the axial loads and the bending loads.

Hence, let the strand be subjected to bending only where the contact is maintained between the outside wires and the center wire. Between any two cross sections along an outside wire, the only way an external load can be applied to the outside wire is along the line of contact. Therefore, if a force per unit length acts on the boundary of a thin wire, of radius R, as shown in Figure 5.1, the following can be written down by transforming the force on the boundary to an equivalent force and couple acting along the centerline of the wire [16]

$$Zds = Z_0 ds_0 \cos\beta - Y_0 ds_0 \sin\beta, \tag{5.1}$$

$$Yds = Z_0 ds_0 \sin\beta + Y_0 ds_0 \cos\beta, \tag{5.2}$$

and, hence,

$$K = 0, \tag{5.3}$$

$$K' = -ZR, \tag{5.4}$$

and

$$\Theta = YR, \tag{5.5}$$

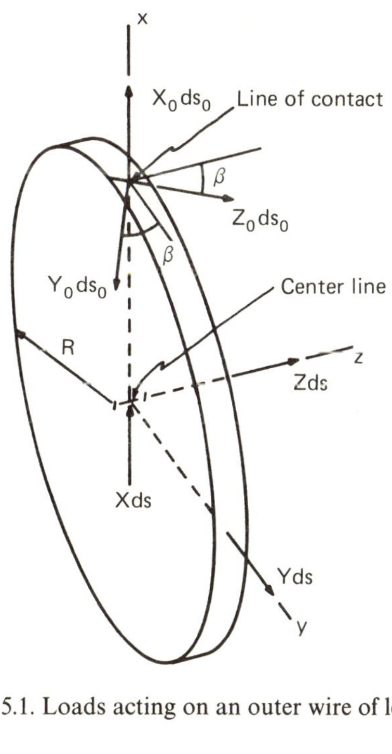

FIGURE 5.1. Loads acting on an outer wire of length *ds*.

where Y_0 and Z_0 are the forces per unit length acting along the contact curve, β is the angle shown in Figure 5.1, ds_0 is a differential length along the contact curve corresponding to a differential length ds along the centerline of an outside wire, and K, K', and Θ are the components of the external moments per unit length along the centerline of an outside wire.

Therefore the differential equations of equilibrium [Eqs. (2.4) through (2.6) and (2.8) through (2.10)] become

$$\frac{dN}{ds} - N'\tau + T\kappa' + X = 0, \tag{5.6}$$

$$\frac{dN'}{ds} - T\kappa + N\tau + Y = 0, \tag{5.7}$$

$$\frac{dT}{ds} - N\kappa' + N'\kappa + Z = 0, \tag{5.8}$$

$$\frac{dG}{ds} - G'\tau + H\kappa' - N' = 0, \tag{5.9}$$

$$\frac{dG'}{ds} - H\kappa + G\tau + N - RZ = 0, \tag{5.10}$$

and

$$\frac{dH}{ds} - G\kappa' + G'\kappa + RY = 0. \tag{5.11}$$

The following interesting observations can be made upon an examination of Eqs. (5.6) through (5.11). Since the original curvatures and twist per unit length of an outside wire are known, assumed values of the deformed curvatures and twist per unit length will yield, by Eq. (2.12), the values of G, G', and H. Equations (5.9) and (5.11) can then be used to solve for N' and Y. Equation (5.10) determines the value of Z in terms of N. Hence, Eqs. (5.7) and (5.8) can now be combined to yield a linear, first-order, ordinary differential equation of T in terms of s. The equations of equilibrium of an outside wire will thus yield the tension T along the wire for assumed curvatures and twist per unit length.

In the deformed configuration, fairly accurate estimates can be made for the deformed curvatures and twist per unit length. The tension T, due to bending in an outside wire, can then be determined from the solution of the differential equation discussed above. This procedure eliminates the difficult task of estimating the axial strain and, hence, T, along an outside wire.

It should be remarked that the above procedure was used to determine the tensions in an outer wire for the axially loaded strand. The final curvatures and twist per unit length were assumed and the loads required to hold the wires in equilibrium were determined from the equations of equilibrium.

Figure 5.2 shows the centerline of an outside wire in a deformed position in which the outside wire is wrapped around the center wire. The deformed components of the curvature and twist per unit length will now be obtained by again projecting the angular velocity vector of a rotating reference frame, which moves with a unit velocity along the deformed centerline of an outside wire. This can be accomplished by noting that the absolute angular velocity vector of the x, y, z frame is equal to the angular velocity vector with respect to the X, Y, Z system plus the angular velocity vector of the X, Y, Z system with respect to the fixed ABC system. If it is assumed that $\tan\alpha = \rho\theta/r\phi$, then

$$\kappa = \frac{\sin\alpha}{\rho}\cos\phi,$$

$$\kappa' = \frac{\cos^2\alpha}{r} - \frac{\sin^2\alpha}{\rho}\sin\phi,$$

and

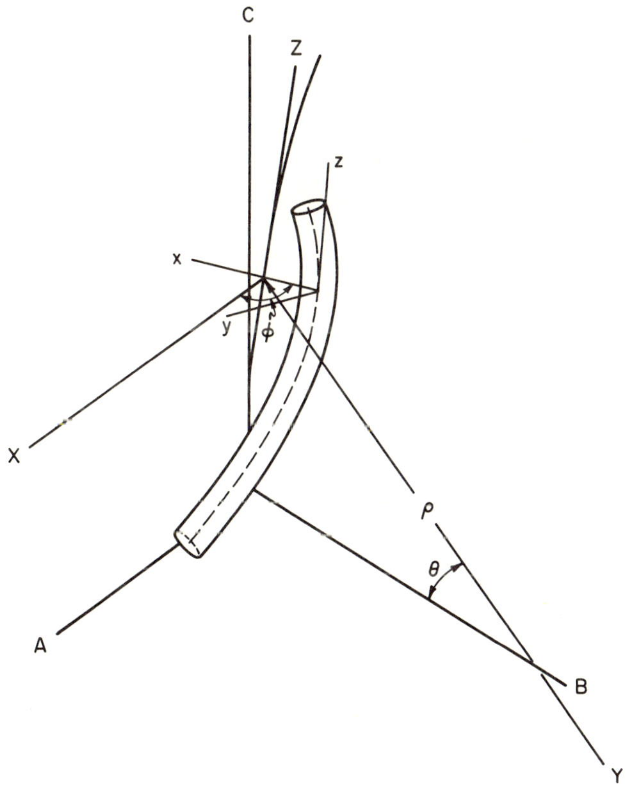

FIGURE 5.2. Helix on a circle.

$$\tau = \frac{\sin\alpha\cos\alpha}{r} + \frac{\sin\alpha\cos\alpha}{\rho}\sin\phi, \tag{5.12}$$

where κ and κ' are the final components of the curvature, τ is the angle of twist per unit length, ρ is the radius of curvature of the center wire, r is the radius to the centerline of an outer wire, α is the helix angle of an outer wire, and ϕ, which lies in the XY plane, is the angle shown in Figure 5.2.

Hence, as mentioned previously, Eqs. (5.9), (5.10), and (5.11) yield

$$N' = \frac{\pi E R^4}{4}\frac{\nu}{(1+\nu)}\frac{\sin\alpha\cos\alpha}{\rho}\left[-\frac{\cos^2\alpha\sin\phi}{r} + \frac{\sin^2\alpha\cos^2\phi}{\rho}\right], \tag{5.13}$$

$$RZ = N + \frac{\pi E R^4}{4}\frac{\nu}{(1+\nu)}\frac{\sin^2\alpha\cos\alpha\sin\phi\cos\phi}{\rho^2}, \tag{5.14}$$

and

$$RY = \frac{\pi E R^4}{4}\frac{\nu}{(1+\nu)}\frac{\sin\alpha\cos^2\alpha\cos\phi}{\rho r}, \tag{5.15}$$

where

$$s = \frac{r}{\cos\alpha}\phi. \tag{5.16}$$

Equation (5.13) results in

$$\frac{dN'}{ds} = \frac{dN'}{d\phi}\frac{d\phi}{ds} = \frac{\pi E R^4}{4}\frac{\nu}{(1+\nu)}\frac{\sin\alpha\cos^2\alpha}{r\rho}\left(-\frac{\cos^2\alpha}{r} + 2\frac{\sin^2\alpha}{\rho}\sin\phi\right)\cos\phi. \tag{5.17}$$

When Eqs. (5.7) and (5.8), (5.13) through (5.15), and (5.17) are combined,

$$\frac{dT}{d\phi} + \frac{r}{\rho}\frac{\left(\frac{r}{R} - \cos^2\alpha + \frac{r}{\rho}\sin^2\alpha\sin\phi\right)}{\cos^2\alpha\left(1 + \frac{r}{\rho}\sin\phi\right)}(\cos\phi)T$$

$$= -\frac{\nu}{1+\nu}\frac{\pi E R^4}{4r^2}\frac{r}{\rho}\left[\frac{r}{R} - \cos^2\alpha + \frac{r}{\rho}\sin^2\alpha\sin\phi\right]$$

$$\times\left[\frac{r}{\rho}\sin^2\alpha\sin\phi - \frac{\left(\frac{r}{R} - \cos^2\alpha + \frac{r}{\rho}2\sin^2\alpha\sin\phi\right)}{\left(1 + \frac{r}{\rho}\sin\phi\right)}\right]\cos\phi. \tag{5.18}$$

Now let $r/\rho \ll 1$. Equation (5.18) then becomes

$$\frac{dT}{d\phi} + \frac{r}{\rho\cos^2\alpha}\left[\frac{r}{R} - \cos^2\alpha\right](\cos\phi)T = \frac{\pi E R^4}{4r^2}\frac{\nu}{(1+\nu)}\frac{r}{\rho}\left[\frac{r}{R} - \cos^2\alpha\right]^2\cos\phi \tag{5.19}$$

and, hence, has the solution

$$T = \frac{C_3}{e^{C_1 \sin \phi}} + \frac{C_2}{C_1}, \tag{5.20}$$

where

$$C_1 = \frac{r}{\rho \cos^2 \alpha}\left[\frac{r}{R} - \cos^2 \alpha\right], \tag{5.21}$$

$$C_2 = \frac{\pi E R^4}{4r^2} \frac{v}{(1 + v)}\left[\frac{r}{\rho} - \cos^2 \alpha\right]^2, \tag{5.22}$$

and C_3 is an arbitrary constant. The constant C_3 will be determined by the equation

$$\int_0^{2\pi} T d\phi = 0, \tag{5.23}$$

which indicates that the change in length due to bending of an outer wire in one lay length is assumed to be zero.

Since $C_1 \ll 1$, Eq. (5.23) yields the value of C_3 as

$$C_3 = -\frac{C_2}{C_1}, \tag{5.24}$$

and, hence, Eq. (5.20) becomes

$$T = \frac{\pi E R^4}{4r^2} \frac{r}{\rho} \frac{v}{(1 + v)}\left[\frac{r}{R} - \cos^2 \alpha\right]^2 \sin \phi. \tag{5.25}$$

Based on the previous discussion, the following can be writte down:

$$G = \frac{\pi E R^4}{4} \frac{\sin \alpha}{\rho} \cos \phi, \tag{5.26}$$

$$G' = -\frac{\pi E R^4}{4} \frac{\sin^2 \alpha}{\rho} \sin \phi, \tag{5.27}$$

$$H = \frac{\pi E R^4}{4(1 + v)} \frac{\sin \alpha \cos \alpha}{\rho} \sin \phi, \tag{5.28}$$

$$Y = \frac{\pi E R^4}{4R\rho r} \frac{v}{(1 + v)} \sin \alpha \cos^2 \alpha \cos \phi, \tag{5.29}$$

$$N' = -\frac{\pi E R^4}{4\rho r} \frac{v}{(1 + v)} \sin \alpha \cos^3 \alpha \sin \phi, \tag{5.30}$$

$$N = -\frac{\pi E R^4}{4\rho} \frac{v}{(1 + v)}\left[\frac{1}{R} - \frac{\cos^2 \alpha}{r}\right] \cos \alpha \cos \phi, \tag{5.31}$$

$$Z = -\frac{\pi E R^4}{4R\rho} \frac{v}{(1 + v)}\left[\frac{1}{R} - \frac{\cos^2 \alpha}{r}\right] \cos \alpha \cos \phi, \tag{5.32}$$

and

$$X = -\frac{\pi E R^4}{4\rho r^2}\frac{v}{(1+v)}\cos^2\alpha\left[\left(\frac{r}{R}\right)^2 + \frac{r}{R}(1 - 2\cos^2\alpha)\right]\sin\phi. \quad (5.33)$$

It should be remembered that the above equations are for bending only when the assumed deformed curvatures and twist per unit length are given by Eq. (5.12).

Equation (5.25) indicates the rather unexpected result that the tension T, although positive for $0 < \phi < \pi$, is small. Let, for example, $R_1 = 0.031$ in., $R_2 = 0.029$ in., $\alpha = 70°$, $\rho = 60$ in., $E = 30 \times 10^6$ psi, and $v = 0.25$. The previous equations yield values of $G = 0.2610\cos\phi$, $G' = -0.2453\sin\phi$, $H = 0.0714\sin\phi$, $X = -10.5874\sin\phi$, $Y = 3.5093\cos\phi$, $Z = -21.3139\cos\phi$, $N = -0.6181\cos\phi$, $N' = -0.0348\sin\phi$, and $T = 3.5277\sin\phi$. Plots of the above loads are shown in Figures 5.3, 5.4, and 5.5.

Now let the total axial load on the strand be 710.3 lb. The contact load for the axial case with no rotation is -196.45 lb/in., and this is sufficient to generate the frictional loads for small coefficients of friction. However, in the axial case, the tensile load in an outside wire is 101.75 lb, which is considerably higher than the maximum tensile load of 3.53 lb due to the bending.

The total stress acting on an outside wire at a point furthest from the so-called neutral axis consists of four parts. The axial and bending stresses

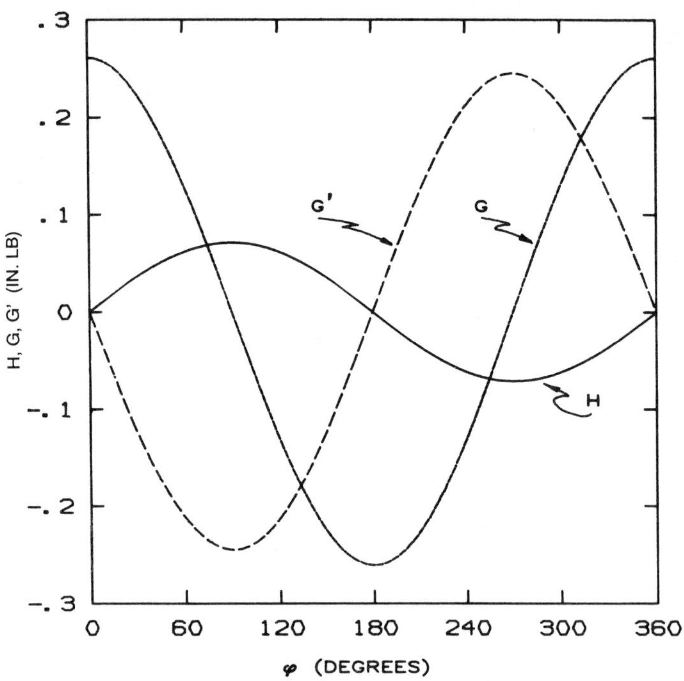

FIGURE 5.3. Bending moments as a functions of ϕ.

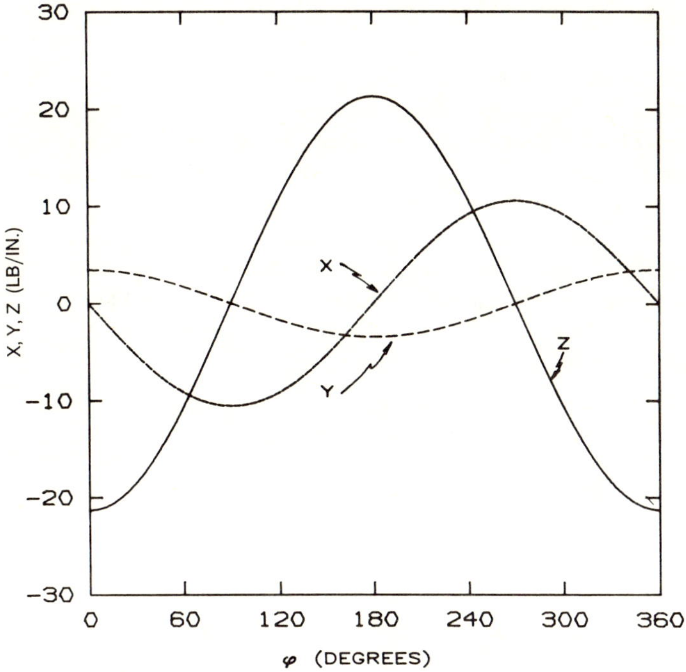

FIGURE 5.4. Contact loads as a function of ϕ.

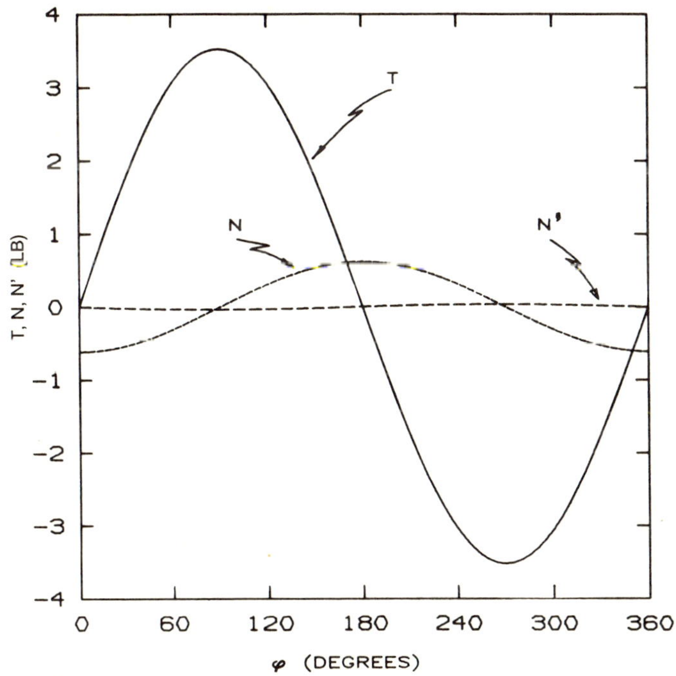

FIGURE 5.5. Loads as a function of ϕ.

due to the axial loading and the axial and bending stresses due to bending with radius ρ. The value of these stresses are 38,510 − 4,950 − 1,330 + 12,800 = 45,040 psi, while the maximum stress on an outside wire at the neutral axis is 38,510 + 13,630 = 52,140 psi. The maximum tensile stress in the center wire is 45,000 + 15,500 = 60,500 psi.

The above results show, for the curvatures and twist assumed, that the tensile stresses, caused by T, due to the bending of a simple strand are small when compared with the tensile stresses due to T for axial loads.

In the case of a spring subjected to pure bending [8], the final curvatures and twist per unit length are approximated by

$$\kappa = \frac{2\sin\alpha}{(2 + \nu\cos^2\alpha)}\frac{1}{\rho}\cos\phi, \tag{5.34}$$

$$\kappa' = \frac{\cos^2\alpha}{r} - \frac{2}{(2 + \nu\cos^2\alpha)}\frac{\sin^2\alpha}{\rho}\sin\phi, \tag{5.35}$$

and

$$\tau = \frac{\sin\alpha\cos\alpha}{r} + \frac{2(1 + \nu)}{(2 + \nu\cos^2\alpha)}\frac{\sin\alpha\cos\alpha}{\rho}\sin\phi. \tag{5.36}$$

The expressions for the curvatures are close to those assumed previously, since $2/(2 + \nu\cos^2\alpha)$ is close to unity. A substitution of the above curvatures and twist per unit length into the equations of equilibrium yields the result:

$$T = 0.$$

Also,

$$N = 0, \quad N' = 0, \quad X = 0, \quad Y = 0, \quad Z = 0, \tag{5.37}$$

$$G = \frac{\pi E R^4}{4}\frac{2\sin\alpha}{(2 + \nu\cos^2\alpha)}\frac{1}{\rho}\cos\phi, \tag{5.38}$$

$$G' = \frac{\pi E R^4}{4}\frac{2\sin^2\alpha}{(2 + \nu\cos^2\alpha)}\frac{1}{\rho}\sin\phi, \tag{5.39}$$

and

$$H = \frac{\pi E R^4}{4}\frac{2}{(2 + \nu\cos^2\alpha)}\frac{\sin\alpha\cos\alpha}{\rho}\sin\phi. \tag{5.40}$$

Consider again the same strand examined previously. In this case, the tensile stress at the so-called maximum distance from the neutral axis is (38,510 − 4,950 + 12,620) = 46,180 psi. The maximum tensile stress acting on an outside wire at the neutral axis is (38,510 + 13,430) = 51,960 psi.

The fact that the tensile loads T, due to bending in the outer wire, are small seems difficult to accept when one examines the bending of a beam in which plane sections remain plane. However, if one considers the pure bending of a spring, the deformed pitch on the side of spring closest to the center of curvature is less than the deformed pitch on the side furthest from the center of curvature and yet the spring does not experience a tensile force.

It is therefore felt that, in the actual case of bending of a simple strand, the tensile forces due to bending are rather small and can be neglected in a determination of the stresses. The bending stresses are generally considerable, however, and should be taken into account. It is also interesting to note that the experiments of Hobbs and Ghavami [17] confirmed that in bending fatigue the first failures occurred in the wires located at the neutral axis in all six tests.

Two estimates for the curvatures and twist per unit length were considered above. In both cases, the tensile forces in an outer wire were very small when compared with the usual tensile loads in the axially loaded case. Since the tensile loads were small, the bending stiffness was predominantly determined by the bending stiffness of each wire, and hence, it is felt that Eq. (3.78) represents a fairly accurate estimate of the bending stiffness and, as mentioned previously, experimental investigations [10] tend to confirm this. It should be noted, however, that small line loads per unit length acting along the contact line due to friction may cause relatively large stresses along the line of contact, since these are line loads.

5.3 Frictional Effects in Wire Rope

In the case of axially loaded ropes, the effects due to friction are again felt to be small, especially in the case in which the outer strands are not in contact. In this case, the strand cross sections are allowed to rotate about an axis passing through the centerline of the rope. In the case of a rope with a fiber core, the cross sections are not as free to rotate about an axis passing through the centerline of the rope and, hence, friction will have some effect.

In the bending and tension of a wire rope with an independent wire rope core, again, since the core tends to keep the strands from contacting each other, the outer strands tend to contact the independent wire rope only and behave somewhat like the simple strand considered in Section 5.2. The total load in a strand due to bending is felt to be small compared with the load for the axially loaded case. Hence, it is felt that the methods employed in Chapters 3 and 4 to determines stresses are fairly accurate for determining stresses in a rope with friction.

5.4 Effective Length of a Broken Center Wire in a Simple Strand

Consider a simple strand axially loaded and not allowed to rotate. Let the center wire stress be σ_1. Now let an imaginary cut be made in the center wire of a simple strand that has a stress σ_0 acting on it. This will divide the center wire into two parts. If $\sigma_0 = \sigma_1$, the parts will not separate. Let the material be frictionless and the axial loads constant. If the stress σ_0 is now reduced, the parts will separate. The outside wires will pick up the additional load being released by the center wire. The contact force between the outer wires and the center wire will increase.

Consider now the same strand in which the material coefficient of friction is infinite. Again, let the center wire be separated into two parts by an imaginary cut and let the stress be σ_0. As stress σ_0 is reduced from stress σ_1, the faces will again separate except at the m_2 contact points. The faces in the center wire will be deformed into a dishlike shape if σ_0 is constant. The outside wires will again pick up the additional load, which is released from the center wire. Since the center wire is not allowed to slip relative to an outside wire, an invocation of Saint-Venant's principle yields a relatively short length in which the stresses will again be distributed as thought the center wire were not cut.

The actual case of the center wire experiencing a fracture is, of course, somewhere between the two cases mentioned above. Let the material have of coefficient of friction μ. As the center wire stress is reduced, two regions may develop in the center wire. The region nearest the fractured end may slip relative to the outside wires and the region that remains will not slip. An estimate of the slip region can be ascertained by using the values of the line contact loads developed in the unbroken center wire strand and the coefficient of friction μ. As mentioned previously, this line load is less than that of the broken center wire strand and thus should give a somewhat conservative result [18].

Let l_s be the length of slip measured from the fractured end of the center wire. A summation of the axial forces on the center wire of length l_s yields (since X_2 is negative)

$$\sigma_s \pi R_1^2 + m_2 \mu X_2 \frac{l_s}{\sin \alpha} = 0, \tag{5.41}$$

in which σ_s is the tensile wire stress at the length l_s. This stress is less than the stress in the unbroken wire. Let l be the length of no slip along the center wire wherein the stress increases from σ_s to nearly, σ_1, the stress in the center wire of an unbroken strand. Thus, the effective length L_e, measured from the fractured end of the center wire, in which the center wire picks up its appropriate share of the load is given by

$$L_e = l_s + l. \tag{5.42}$$

It should be noted that the maximum slip length can be obtained by setting $\sigma_s = \sigma_1$ in Eq. (5.41). Let this length be denoted l_{ms}, and thus,

$$\sigma_1 \pi R_1^2 + m_2 \mu \frac{X_2 l_{ms}}{\sin \alpha_2} = 0. \tag{5.43}$$

Equation (5.43) indicates, the important result, that, for $\tau = 0$, this maximum slip length, l_{ms}, is independent of the axial load F, since σ_1 and X_2 are both linear in F.

Example 5.1. Consider a simple strand with the following properties: $R_1 = 0.03155$ in., $R_2 = 0.02893$ in., $P_2 = 1.3$ in., $v = 0.25$, $E = 29 \times 10^6$ psi, $m_2 = 6$, and $\mu = 0.1$. Let $\xi_1 = 0.003$ and $\tau = 0$. The theory in Chapter 3 yields a value of $X_2 = -266.9$ lb./in. Equation (5.43) becomes

$$29 \times 10^6 \times 0.003 \times \pi \times (0.03155)^2 + \frac{6 \times 0.1 \times -266.9 \times l_{ms}}{0.95983} = 0,$$

with the result that $l_{ms} = 1.631$ in. The actual slip length will be less than this value. The actual stress in an unbroken center wire for the above strains is 87,000 psi. If for example the actual slip length was 1.378 in., Eq. (5.41) would yield a value of $\sigma_s = 73{,}530$ psi. The remainder of the wire would not slip and, thus, in a very short length, the stress in the center wire will nearly be 87,000 psi.

5.5 Effective Length of a Broken Outer Wire in a Rope

Figure 5.6 shows the fatigue failure of an independent wire rope core of 0.875-in. diameter regular lay 6 × 25 IWRC wire rope subjected to axial loading only. The outer wires show no visable signs of fracture, although the wires in the internal core of the rope are fractured into relatively short lengths. The wires in the internal core are subjected to the largest axial stresses when the rope is subjected to axial loading only. The fractures of the short lengths appear to be the result of friction, which enables the center wires to sustain their appropriate share of the load in a relatively short length. If the wires were very smooth, the tensile stress in the wires would remain at low levels for a relatively long length along the wires from the fractured end and thus would not fracture into small pieces.

In general, the performance of a wire rope is usually not affected by an occasional broken wire in a rope. In fact in long ropes, it is possible to have every wire broken somewhere along its length and yet the rope will still function satisfactorily. There is sufficient friction in the rope to enable a broken wire to develop its total share of the load in a relatively short length. The problem, however, becomes acute when a number of broken wires occur in a relatively short length along the rope. In the mining industry, for example, some ropes may measure up to 7,000 ft. Broken wires in the rope are monitored by passing the rope through an electromagnetic field that will detect broken wires. Many times the number of broken wires in a given length is used as a retirement criterion for the rope. Since visual inspection will detect only broken outer wires and since the outer strands carry the main portion of the load, an estimate of the effective length of an outer wire is important.

FIGURE 5.6. Axial fatigue failure of an independent wire rope core of a 6 × 25 IWRC.

The method employed to determine the effective length of a center wire in a simple strand can now be extended to the complex cross sections of wire rope. Consider the 6 × 19 Seale IWRC rope shown in Figure 1.1. In Chapter 4, equations are presented that will determine the contact force per unit length X_3. Since there are nine wires in an outer Seale strand, each outer wire experiences a contact load, F_c, in which (since X_3 is negative)

$$F_c = -\frac{X_3 p_{33}}{9}, \tag{5.44}$$

where p_{33} is the pitch of the outer wires in the strand. Each time an outer wire comes in contact with the wire rope core it is subjected to the load F_c acting on both sides of the wire. Thus, if a broken wire is slipping relative to the other adjacent wires at the contact point, the change in tension T_A of the broken wire across the contact point will be given by

$$T_A = \mu 2F_c. \tag{5.45}$$

Also, as the outer wire wraps around the strand it experiences line contact loads with adjacent wires. This line contact load increases as the tension in an outer wire increases, and thus, if the wire is slipping, its tension will increase due to the friction caused by the line contact loads.

Again the line contact load X_{33} that acts on an outer unbroken wire as it wraps around the strand can be calculated by the methods developed in Chapter 4. When the wire is fractured, the contact line load decreases. It will now be assumed that the line contact load in the broken outer wire is proportional to the tension in the wire. When the tension in the broken wire reaches the tension in an unbroken wire, the line contact load will be the same as that of the unbroken wire. It would be noted that the line contact load X_{33}, acting on an outer wire, is the resultant of the two line contact loads of the inner lay of wires in an outside strand (see Figure 4.3). Each outside wire comes in contact with two inner wires. This tends to increase the frictional force, since the components of X_{33} must be used. Thus, the frictional force per unit length f, due to the resultant contact load X_{33}, is (since X_{33} is negative)

$$f = -\mu \frac{X_{33}}{\cos \theta_{33}}, \tag{5.46}$$

in which θ_{33} is the angle on a given cross section of an outer wire between the resultant line load X_{33} and a point of contact.

Example 5.2. Consider the 1.306-in. diameter, Seale rope used in Section 4.2. Let the axial strain be 0.003 and the rotational strain be zero. From the theory developed in Chapter 4, the following values are obtained: $X_{3*} = -1{,}725$ lb/in., $X_{33} = 436.8$ lb/in., $T_{33} = 510$ lb, $\theta_{33} = 21.99°$, $p_{33} = 2.467$ in. and $S_{33} =$ 2.646 in., where T_{33} is the tension in an outer unbroken wire and S_{33} is the length of an outer wire from contact point to contact point. Let an outer wire be fractured at a contact point. Thus, for a length of 2.646 in the tension in this outer wire will be zero. At the next contact point, the contact load, given by Eq. (5.44), is $F_c = 472.8$ lb; thus, Eq. (5.45) yields an increase in tension, T_A, in the outer wire in which $T_A = 0.1 \times 2 \times 472.8 = 94.56$ lb, where $\mu = 0.1$ has been used. Thus, as mentioned previously, the contact line load X_{33}, in the next section from contact point to contact point, will be assumed as $-436.8 \times 94.56/510 = -80.99$ lb/in. Equation (5.46) yields $f = 0.1 \times 80.99/0.9273 =$ 8.733 lb/in. This frictional force acts on the next length of 2.646 in., and thus the tension in the outer broken wire just after the next contact point becomes $94.56 + 8.733 \times 2.646 = 212.2$ lb. This process is continued until the tension in the wire is at least 510 lb (the tension in an unbroken wire).

In this example, after a length of just greater than $4 \times 2.646 = 10.58$ in., the calculated tension turns out to be 540.8 lb, which is greater than the 510 lb. Thus, the maximum effective length of this broken wire will be taken as 10.58 in. This translates into a length of 9.284 in. along the rope. It is interesting to note that the pitch of an outer strand is 7.88 in., and thus the effective length of a broken outer wire is about 18% greater than the pitch.

6
Testing of a Wire Rope

6.1 Axial Testing of a Wire Rope

Initially, a load deformation curve was obtained experimentally for a 1.306-in. diameter, 6 × 19 Seale IWRC rope on a 600,000-lb testing machine. A plot of the load as a function of the deformation is shown in Figure 6.1. In this test, a 45-in. gage length was used and the total cross-sectional area $\Sigma \pi R_i^2 = 0.727$ in.^{2}. The experimental effective modulus of the rope is about 18,400,000 psi, although the preceding theory predicts an effective modulus, based on $E =$ 29,000,000 psi, of about $0.7 \times E = 20{,}300{,}000$ psi. The theoretically determined modulus is higher than the experimentally determined one, since contact deformation between the wires is neglected; also, the outer strands tend to settle inside the radius of the IWRC, since the line load on the outer strand acts only at the contact points. This last condition is especially apparent in the initial loading of a rope. In the above theory it was assumed that the modulus of elasticity of the wire material was 29,000,000 psi.

Additional tests [15] were run on an $\frac{1}{2}$-in. 6 × 25 IWRC wire rope to determine the modulus of elasticity of the wire material and to determine the effective modulus of a strand in the rope. A clip gage with a nominal gage length of 1 in. was used to determine the strain in an individual wire, as shown in Figure 6.2. A drop of cyanoacrylate cement was applied at each contact point to prevent slipping on the specimen. The specimen was loaded by file grips, and fracture of the wire invariably occurred at a grip. The actual diameter of each wire was measured with a micrometer calipter, and the data for the wire diameters, reported in Table 6.1, are felt to be accurate to within 0.0002 in. on the basis of a statistical analysis of several measured values of the wire diameter for a given wire. Also reported in Table 6.1 are the corresponding spring rates for the various wires and the resulting, experimentally determined values of Young's modulus for the wires. The spring rate $\Delta F/\Delta\varepsilon$ was determined by taking the slopes of one or more unloading/reloading portions of the load-deflection curve, such as portions 2 and 3 of the curve in Figure 6.3. Only those curves unloading/reloading curves for which the offset yield strain was less than 0.3% were used.

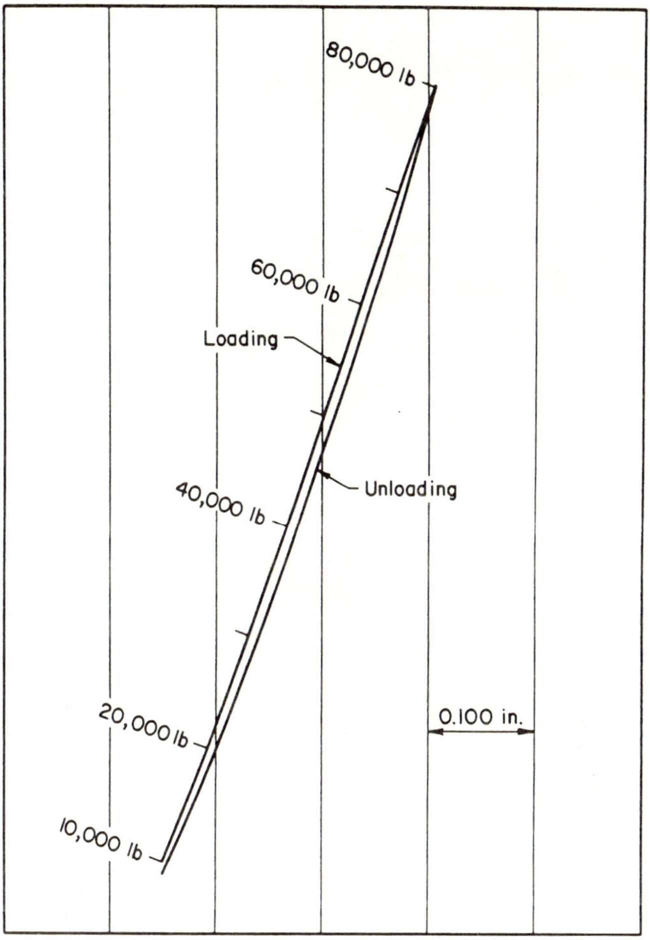

FIGURE 6.1. Load deformation curve of a 6 × 19 Seale IWRC.

On the basis of the results in Table 6.1, it can be concluded that the Young's modulus of the steel is about 29.6×10^6 psi. The 0.2%-offset yield strength of the wire material was observed to be about $280 \times 10^3 \pm 30 \times 10^3$ psi. The ultimate strength was not accurately determined because failure never occurred away from the grips; however, it can be stated that the ultimate strength exceeds the 0.2%-offset yield strength by at least 5%.

Specimens of initially straight strand formed of a 0.037-in. core wire wrapped left-handed, in turn, by six 0.035-in. wires, six 0.015-in. filler wires, and twelve 0.032-in. wires, were tested in a Richle 200,000-lb tension/compression machine as shown in Figure 6.4. A 12-in., beam-type clip gage was used to determine the axial strain in the specimen. Four-inch-long steel collars with zinc-filled tapered internal bores were used to load the specimen; the individ-

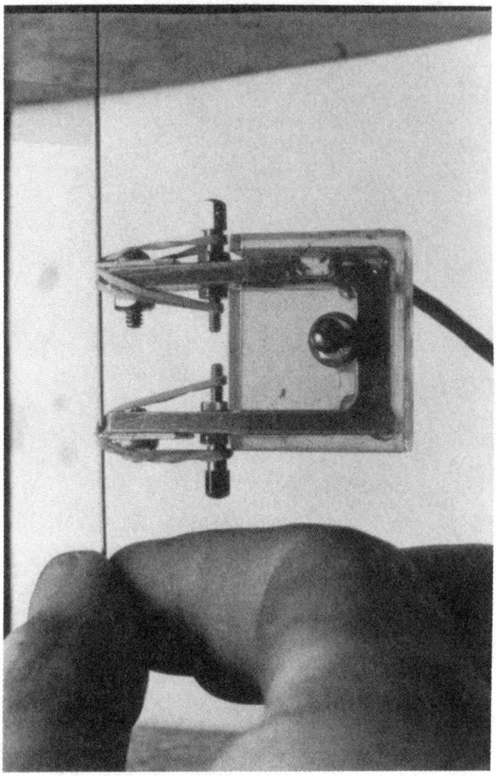

FIGURE 6.2. Clip gage.

TABLE 6.1. Young's modulus.

Nominal wire diameter (in.)	Actual wire diameter (in.)	Actual cross-sectional area (in.^{2})	Spring rate $\Delta F/\Delta\varepsilon$ (lb)	Young's modulus E (psi)
0.015	0.0149	0.1744×10^{-3}	5,240	30.0×10^6
			5,190	29.8×10^6
0.032	0.0322	0.8143×10^{-3}	22,500	27.6×10^6
			24,900	30.6×10^6
0.035	0.0349	0.95656×10^{-3}	28,900	30.2×10^6
			29,200	30.5×10^6
0.037	0.0379	1.128×10^{-3}	32,500	28.8×10^6

ual wires within the strand had been splayed in the collar prior to the pouring of the molten zinc. In the testing machine, the collars were rigidly held by means of *V*-groove grips, as shown in Figure 6.5. In view of the massiveness of the testing machince and the method of gripping that was employed, one can argue that the ends of the specimen were effectively prevented from rotating.

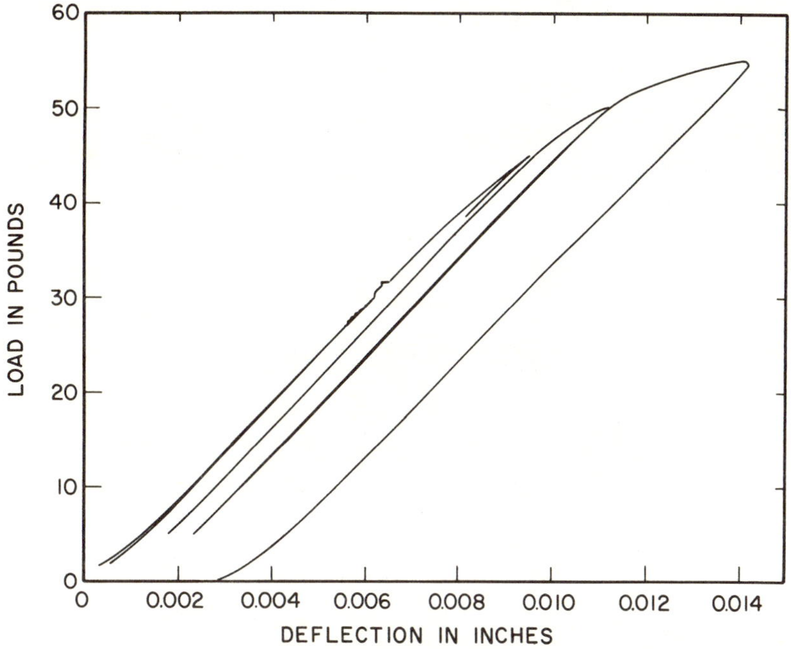

FIGURE 6.3. Load deflection curve.

FIGURE 6.4. 200,000-lb testing machine.

FIGURE 6.5. Strand rope end connection.

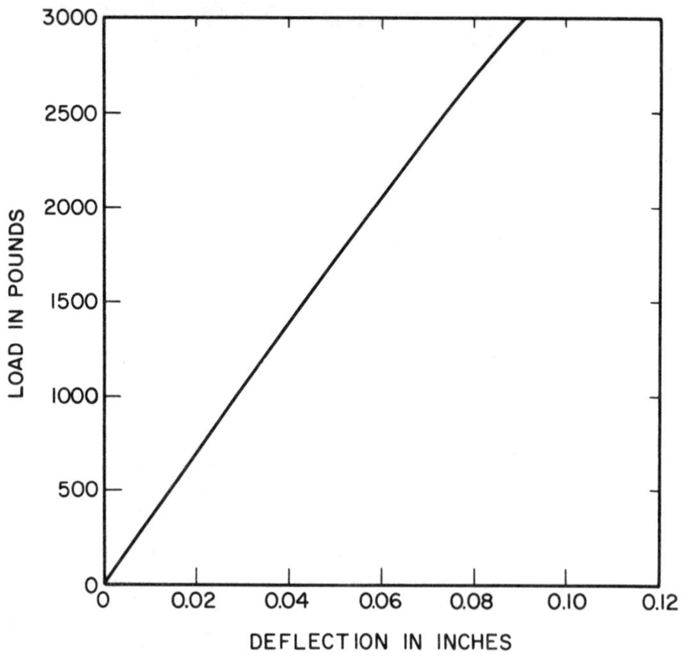

FIGURE 6.6. Load deflection curve for a strand.

A typical load-deflection curve for the strand is shown in Figure 6.6. In the test illustrated, the loading curve is slightly nonlinear right from the origin. The spring rate $\Delta F/\Delta\varepsilon$ near the origin is approximately 0.42×10^6 lb. The metallic area of the strand A_s is 0.01768 in.^{2}. The effective modulus is then given by $E_e = \Delta F/A_s\Delta\varepsilon = 0.42 \times 10^6/0.01768 = 24 \times 10^6$ psi. It should be noted that the computed value of E_e is about 81% of the value of E of the wire material.

A regular-lay $\frac{1}{2}$-in. diameter, 6×25 *F* IWRC wire rope made from six of the left-lay strands just described was tested. The strands were wrapped right-handed about a core, which is itself a right-lay, lang lay 6×7 IWRC, the smaller IWRC being a seven wire, right-lay strand. The total metallic area of the IWRC was computed to be 0.026 in.^{2}. The total metallic area A of the rope was computed to be $A = 0.132$ in.^{2}.

Tests on the $\frac{1}{2}$-in. diameter rope were run on the same Richle 200,000-lb machine used to test the strands. Also, the same 12-in., gage-length clip gage was used to determine the axial deflection. The ends of the rope specimens were held by zinc-filled cylindrical sockets that were gripped by *V*-groove grips to prevent rotation.

The detail of the contact between the clip gage and the rope is shown in Figure 6.7. The clip gage spanned approximately 22 crests of the six-strand

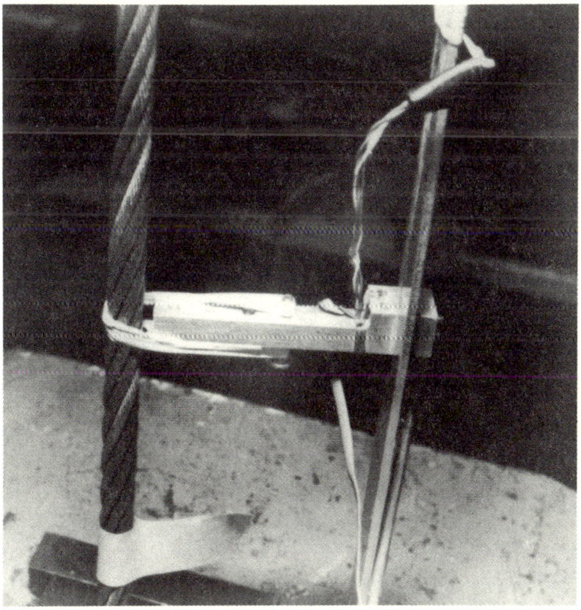

FIGURE 6.7. Clip gage attached to wire rope.

rope, which means that the ends of the clip gage were not in contact with the same strand within the rope. A plot of the load-deflection curve for the rope is shown in Figure 6.8. Along the initial portion of the load-deflection curve of the virgin rope, the spring rate is rather small. Along the linear region along the second unloading curve (between 20×10^3 lb and 10×10^3 lb), $\Delta F/\Delta \varepsilon$ is approximately 2.3×10^6 lb. The effective Young's modulus E_e for the rope is then given by the equation $E_e = \Delta F/A\Delta\varepsilon = 2.3 \times 10^6/0.132 =$ 17,400,000 psi.

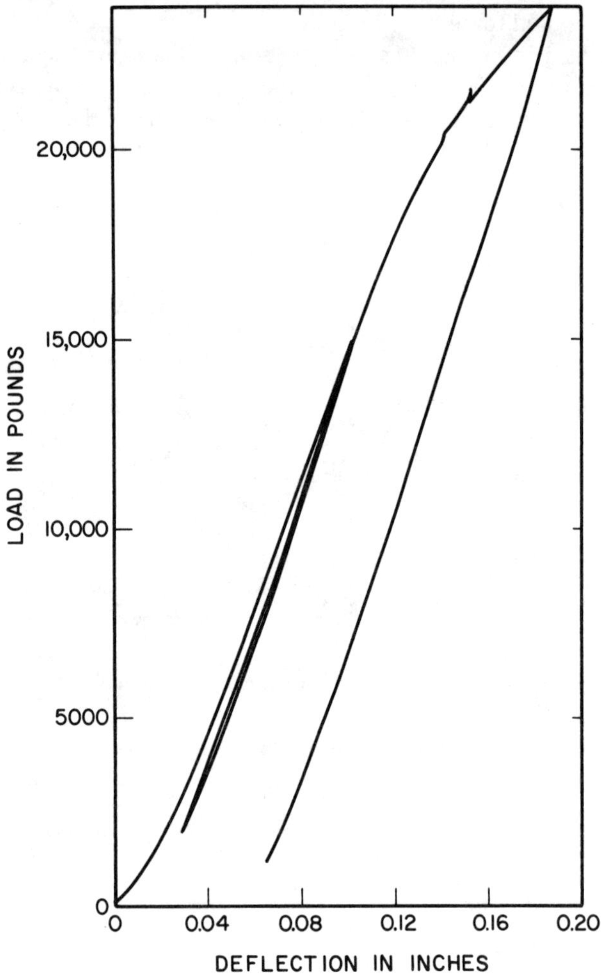

FIGURE 6.8. Load deflection curve for rope.

6.2 Effect of Rope Size on Rope Strength

Let a group of straight wire specimens of different sizes be considered. There is a higher probability for sizable flaws to exist in the larger specimens simply because they have a greater volume and a greater surface area and because high localized stresses may occur near the flaws. Also, it is noted that straight wire is the product of a series of drawing operations and this cold-working increases the tensile strength of the material. Therefore larger-diameter wire generally exhibits lower unit fracture strength because of the higher probability of containing sizable flaws and a smaller degree of cold-working. The following example will show that larger-diameter ropes exhibit lower unit nominal strength than smaller-diameter ropes.

Let, for example, the nominal strength S of a wire rope be

$$S = f_1(d, v, E, n, K, \text{type of rope}), \tag{6.1}$$

where d is the diameter of the rope, v is Poisson's ratio of the stock material, E is the modulus of elasticity of the stock material, n is the strain hardening coefficient of the stock material, and K is a stock material constant, which is defined as the true stress at a true strain of 1.0. The term *stock material* is defined as the raw material for manufacturing the wire. Other properties of the stock material could be included in Eq. (6.1). They will not, however, affect the results pointed out below.

A dimensional analysis [19] of Eq. (6.1) yields

$$S = Ed^2 f_2\left(v, n, \frac{K}{E}, \quad \text{type of rope}\right). \tag{6.2}$$

Therefore, for the same type of rope and the same stock material, Eq. (6.2) indicates that

$$S = CEd^2, \tag{6.3}$$

where C is a constant. Equation (6.3) indicates that the value of S of a wire rope should increase with the diameter squared. Therefore

$$S_p = S_m\left(\frac{d_p}{d_m}\right)^2, \tag{6.4}$$

where the subscripts p and m refer to the prototype and model.

The actual nomial strengths and calculated nominal strengths for two types of ropes are given in Table 6.2, where the actual nominal strengths were taken from the *Yellow Strand Wire Rope Handbook* [20]. The calculated nominal strengths were calculated according to Eq. (6.4) and were based on the $\frac{1}{2}$-in. diameter wire rope, that is, $S_m = 23{,}000$ lb when $d_m = \frac{1}{2}$ in. The values of the percent difference were calculated by the following equation

$$\text{Percent difference} = \frac{\text{Calculated } S\text{-Actual } S}{\text{Actual } S} \times 100. \tag{6.5}$$

TABLE 6.2. Actual and calculated nominal strengths for two types of rope.

	6 × 19, 6 × 37, IWRC			6 × 25, 6 × 30, Flattened Strand fiber core		
d_p (in.)	Actual S (lb)	Calculated S (lb)	Difference (%)	Actual S (lb)	Calculated S (lb)	Difference (%)
$\frac{1}{4}$	5,880	5,750	−2.2			
$\frac{3}{8}$	13,120	12,994	−1.4			
$\frac{1}{2}$	23,000	23,000	0.0	23,600	23,600	0.0
$\frac{5}{8}$	35,800	35,940	0.4	36,600	36,870	0.7
$\frac{3}{4}$	51,200	51,750	1.1	52,400	53,100	1.3
$\frac{7}{8}$	69,200	70,440	1.8	70,800	72,270	2.1
1	89,800	92,000	2.4	92,000	94,400	2.6
$1\frac{1}{4}$	138,800	143,750	3.6	142,000	147,500	3.9
$1\frac{1}{2}$	197,800	207,000	4.6	202,000	212,400	5.1
$1\frac{3}{4}$	266,000	281,750	5.9	272,000	239,100	6.3
2	344,000	368,000	7.0	352,000	377,600	7.3
$2\frac{1}{2}$	524,000	575,000	9.7			
3	740,000	828,000	11.9			
$3\frac{1}{2}$	982,000	1,127,000	14.8			
4	1,254,000	1,472,000	17.4			

Also, all the ropes were made of the same material—Yellow Strand Improved Plow Steel.

It should be noted that the higher the value of d_p/d_m, the higher the percent difference, which indicates that the larger-diameter ropes exhibit lower unit nominal strengths than the smaller-diameter ropes. Also, the percent difference seems to be independent of the type of rope [21].

Now let the size effect be taken into account. The nominal strength S for a given type of rope can be assumed as

$$S = f_3(d, v, E, n, K, d_0), \tag{6.6}$$

where d_0 is a specified reference rope diameter and is introduced for the purpose of describing the size effect. By dimensional analysis, Eq. (6.6) becomes

$$\frac{S}{Ed^2} = f_4\left(\frac{d}{d_0}, v, h, \frac{K}{E}\right). \tag{6.7}$$

For given stock material, Eq. (6.7) can be written as

$$\frac{S}{Ed^2} = \frac{S_0}{Ed_0^2} f_5\left(\frac{d}{d_0}\right), \tag{6.8}$$

since S_0/Ed_0^2 is a constant, and S_0 is the actual nominal strength for the rope of diameter d_0.

Let, for example, d_0 be chosen as $\frac{1}{2}$ in. Figure 6.9 shows a plot of the function f_5 for a 6 × 19 IWRC wire rope. Again all the data were taken

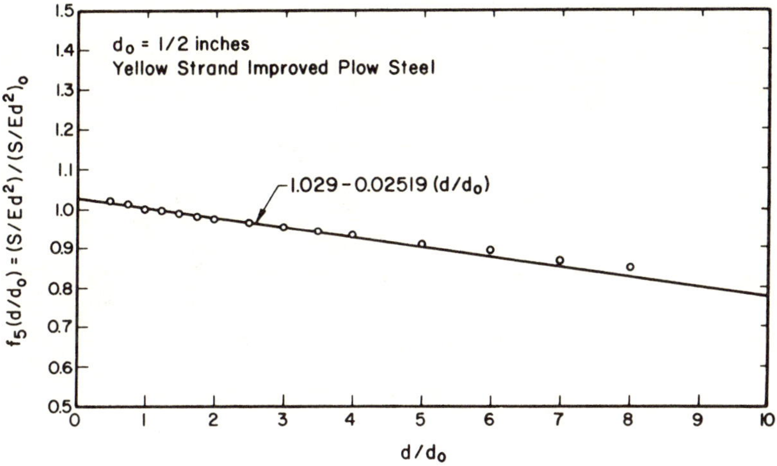

FIGURE 6.9. Plot of f_5 as a function of d/d_0.

TABLE 6.3. Nominal strengths for a 6 × 25 Flatten-Strand fiber-core wire rope.

d (in.)	d/d_0	Actual S (lb)	Calculated S (lb)	Difference (%)
$\frac{1}{2}$	1.00	23,600	23,686	0.36
$\frac{5}{8}$	1.25	36,600	36,778	0.49
$\frac{3}{4}$	1.50	52,400	52,626	0.43
$\frac{7}{8}$	1.75	70,800	71,174	0.53
1	2.00	92,000	92,368	0.40
$1\frac{1}{4}$	2.50	142,000	142,467	0.33
$1\frac{1}{2}$	3.00	202,000	202,477	0.24
$1\frac{3}{4}$	3.50	272,000	271,952	−0.02
2	4.00	352,000	350,477	−0.44

from the *Yellow Strand Wire Rope Handbook*. When the technique of linear regression is used, f_5 can be expressed approximately as

$$f_5 = 1.029 - 0.02519\frac{d}{d_0}, \tag{6.9}$$

where $d_0 = \frac{1}{2}$ in.

For a given stock material, Eqs. (6.8) and (6.9) are valid for any reasonable type of rope. Equations (6.8) and (6.9) can now be applied to a 6 × 25 Flatten-Strand fiber-core wire rope. The results are shown in Table 6.3. Table 6.3 indicates that Eqs. (6.8) and (6.9) yield very accurate predictions for the nominal strengths. It therefore seems reasonable to consider d_0 as one of the primary parameters in the testing of wire rope.

6.3 Effect of Rope Size on Fatigue Life

Consider now the fatigue testing of a wire rope running over a sheave. In this section, a testing methodology for small-diameter ropes will be introduced such that the fatigue life of larger-diameter ropes can be ascertained from the testing of small-diameter ropes. Again, before a testing program is undertaken, a dimensional analysis of the problem should be made.

In order to simplify the problem, only the effects of the axial tension in the rope T, the sheave diameter D, and the rope diameter d on the fatigue life of the rope will be studied. It will now be assumed that the number of cycles to failure N_f for a given type of rope can be assumed to be

$$N_f = f_6(T, D, d, E, \nu, \varepsilon_f, c, \sigma'_f, b, d_0), \tag{6.10}$$

where ε'_f is the fatigue ductility coefficient of the stock material, c is the fatigue ductility exponent of the stock material, σ'_f is the fatigue strength coefficient of the stock material, and b is the fatigue strength exponent of the stock material. Again, additional parameters may be listed but they will not affect the conclusions drawn below. Also d_0 is a specified reference rope diameter, which is introduced for the purpose of describing the size effect.

A dimensional analysis of Eq. (6.10) results in

$$N_f = f_7\left(\frac{T}{Ed^2}, \frac{D}{d}, \varepsilon'_f, c, \frac{\sigma'_f}{E}, b, \nu, \frac{d}{d_0}\right). \tag{6.11}$$

If the tests are now limited to a specific stock material, Eq. (6.11) becomes

$$N_f = f_8\left(\frac{T}{Ed^2}, \frac{D}{d}, \frac{d}{d_0}\right). \tag{6.12}$$

The following test procedure may now be followed in order to determine the size effect on the fatigue life.

1. Choose a value of d_0.
2. Let $d/d_0 = a_1$, where a_1 is a number determined by the diameter d of the rope being tested.
3. Let $D/d = b_1$, where b_1 is a number determined by the diameter D of the testing sheave.
4. Plot the N_f versus T/Ed^2 curve by changing the values of T.
5. Let $D/d = b_2, b_3, \ldots$, and repeat step 4. A family of N_f versus T/Ed^2 curves for $D/d = b_1, b_2, \ldots$, and $d/d_0 = a_1$ can be constructed.
6. Change the value of d/d_0 by changing d and repeat the above procedure. Finally, the effects of T/Ed^2, D/d, and d/d_0 on N_f for this type of rope can be determined by using extrapolation and interpolation.

An example will be given for the purpose of describing the foregoing procedures. The experimental data shown in Table 6.4 are adopted from references 22 and 23. The type of rope under consideration is a 6 × 41—FWS-Lang Lay-IWRC wire rope for which it is assumed that $E = 29 \times 10^6$ psi. Each N_f in Table 6.4 is an average of two tests; β, which is sometimes used to predict the fatigue life, is called the Drucker–Tachau bearing pressure ratio, which is defined [24] as

$$\beta = \frac{2T}{UDd}, \tag{6.13}$$

where U is the ultimate strength of the wire material. The purpose of this procedure is to predict the fatigue life of 3-in. diameter rope from the test data of the 0.75-in. and 1.5-in. diameter rope. The error is about 100% if one uses the Drucker–Tachau criterion.

Let the values of d_0 be chosen as 0.75 in. Figure 6.10 shows the relationship

TABLE 6.4. Fatigue life data.

d(in.)	D(in.)	T(Kips)	T/Ed^2	D/d	d/d_0	N_f(cycles)	β
0.75	15.0	17.83	1.093×10^{-3}	20	1	74,665	11.10×10^{-3}
0.75	15.0	13.37	0.820×10^{-3}	20	1	101,985	8.35×10^{-3}
0.75	22.5	17.83	1.093×10^{-3}	30	1	123,200	7.42×10^{-3}
0.75	22.5	13.37	0.820×10^{-3}	30	1	183,200	5.57×10^{-3}
1.50	30.0	59.06	0.905×10^{-3}	20	2	58,671	9.16×10^{-3}
1.50	30.0	53.60	0.821×10^{-3}	20	2	65,737	8.33×10^{-3}
1.50	45.0	70.29	1.077×10^{-3}	30	2	95,511	7.29×10^{-3}
1.50	45.0	53.60	0.821×10^{-3}	30	2	131,339	5.56×10^{-3}
3.00	90.0	330.00	1.26×10^{-3}	30	4	29,346	9.40×10^{-3}

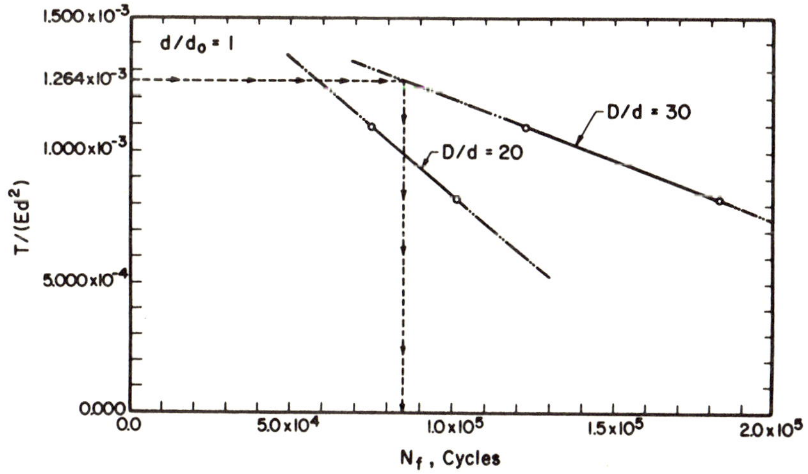

FIGURE 6.10. Plot of N_f as a function of T/Ed^2 and D/d for $d/d_0 = 1$.

between T/Ed^2, D/d, and N_f for $d/d_0 = 1$. Figure 6.11 shows the same relationships for $d/d_0 = 2$. In this example, it is assumed that the relationship between T/Ed^2 and N_f is linear, since only two sets of data were provided. However, it should be noted that more than two sets of data should be provided for one curve in practical cases and then the relationship would probably not be linear.

Consider now the prediction of the bending fatigue life of a 3-in. diameter wire rope running over a 90-in. diameter sheave, where the rope is subjected to an axial tension of 330 kips. Hence $D/d = 30$ and $T/Ed^2 = 1.264 \times 10^{-3}$.

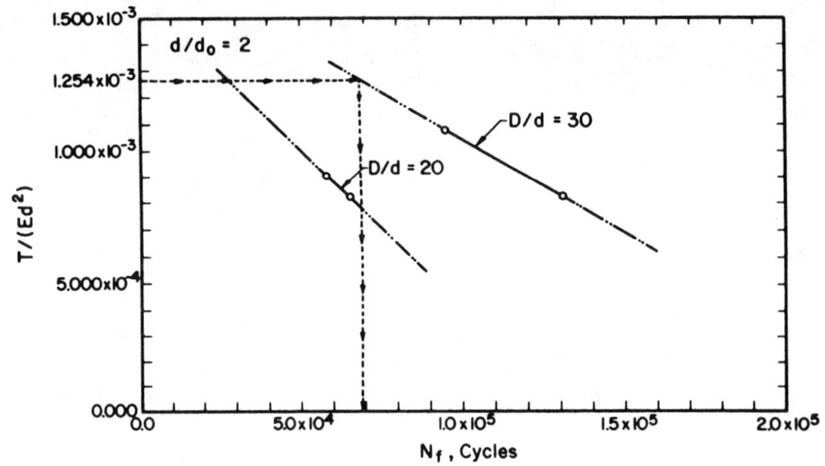

FIGURE 6.11. Plot of N_f as a function of T/Ed^2 and D/d for $d/d_0 = 2$.

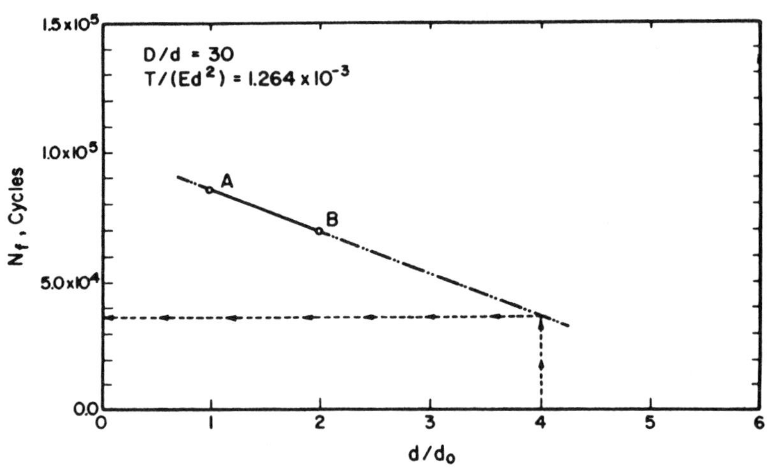

FIGURE 6.12. Plot of fatigue life prediction as a function of d/d_0.

The value of N_f for $D/d = 30$, $T/Ed^2 = 1.264 \times 10^{-3}$, and $d/d_0 = 1$ can be obtained from Figure 6.10, and hence, $N_f = 85{,}600$ cycles. Also when $d/d_0 = 2$, Figure 6.11 yields $N_f = 69{,}340$ cycles. Hence, the points A and B are shown plotted in Figure 6.12. This yields a plot of N_f as a function of d/d_0 for $D/d = 30$ and $T/Ed^2 = 1.264 \times 10^{-3}$. Finally, the value of N_f for the 3-in. diameter rope, that is $d/d_0 = 4$, can be ascertained by utilizing the technique of extrapolation; hence, $N_f = 36{,}780$ cycles. Since the test result of this example is 29,346 cycles, this yields an error of about 25%.

7
Birdcaging in Wire Rope

7.1 Equations of Motion

In many cases, wire rope is subjected to impact loads which send axial and torsional responses up and down the rope. If the loads are severe enough, the outer strands can separate from the core in a permanent manner and thus render the rope useless. Such a phenomenon is generally known as birdcaging. A bird cage is a term often used to describe the permanent appearance of a wire rope forced into compression. Figure 7.1, taken from the *Wire Rope Users Manual* [4], depicts such an event.

Consider again the simple straight strand shown in Figure 3.1. The following can be written down by combining several equations in Chapter 3.

$$\xi_2 = a_1\varepsilon + a_2\beta \qquad \Delta\alpha_2 = a_3\varepsilon + a_4\beta \tag{7.1}$$

$$\Delta\kappa_2 = a_5\xi_2 + a_6\Delta\alpha_2 \qquad \Delta\tau_2 = a_7\xi_2 + a_8\Delta\alpha_2 \tag{7.2}$$

$$\begin{aligned} a_1 &= \frac{r_2\tan^2\alpha_2 - \nu R_1}{r_2 + r_2\tan^2\alpha_2 + \nu R_2} & a_2 &= \frac{r_2^2\tan^2\alpha_2}{R(r_2 + r_2\tan^2\alpha_2 + \nu R_2)} \\ a_3 &= \frac{(1+\nu)r_2\tan\alpha_2}{(r_2 + r_2\tan^2\alpha_2 + \nu R_2)} & a_4 &= -\frac{r_2^2\tan^2\alpha_2}{R(r_2 + r_2\tan^2\alpha_2 + \nu R_2)} \\ a_5 &= \frac{\nu\cos^2\alpha_2}{r_2} & a_6 &= \left(\frac{\nu R_1\cos\alpha_2}{r_2\tan\alpha_2} - 2\sin\alpha_2\right)\frac{\cos\alpha_2}{r_2} \end{aligned} \tag{7.3}$$

$$a_7 = \frac{\nu\sin\alpha_2\cos\alpha_2}{r_2} \quad \text{and} \quad a_8 = \frac{1 - 2\sin^2\alpha_2}{r_2} + \frac{\nu R_1\cos^2\alpha_2}{r_2^2},$$

where, again, ε is the axial strain, β is the rotational strain, ν is Poisson's ratio, ξ_2 is the axial wire strain in an outer wire, $\Delta\alpha_2$ is the change in helix angle of an outer wire, $\Delta\kappa_2$ is the change in curvature of an outer wire, and $\Delta\tau_2$ is the twist per unit length of an outer wire. The expressions for the bending moment G'_2 and the twisting moment H_2 in an outer wire are

$$G_2' = \frac{\pi E R_2^4}{4}(k_1\varepsilon + k_2\beta) \quad \text{and} \quad H_2 = \frac{\pi E R_2^4}{4(1+\nu)}(k_3\varepsilon + k_4\beta), \tag{7.4}$$

FIGURE 7.1. A bird cage in a wire rope.

where E is the modulus of elasticity and

$$k_1 = a_1a_5 + a_3a_6, \quad k_2 = a_2a_5 + a_4a_6,$$
$$k_3 = a_1a_7 + a_3a_8, \quad \text{and} \quad k_4 = a_2a_7 + a_4a_8.$$

Also

$$N'_2 = H_2\frac{\cos^2\alpha_2}{r_2} - G'_2\frac{\sin\alpha_2\cos\alpha_2}{r_2}, \tag{7.5}$$

$$T_2 = \pi ER_2^2\xi_2 = \pi ER_2^2(a_1\varepsilon + a_2\beta), \tag{7.6}$$

and

$$X_2 = N'_2\frac{\sin\alpha_2\cos\alpha_2}{r_2} - T_2\frac{\cos^2\alpha_2}{r_2}, \tag{7.7}$$

where N'_2 is the shearing force in an outer wire, T_2 is the tensile force in an outer wire, and X_2 is the contact line load per unit length acting on an outer wire.

Since G'_2 and H_2 are functions of ε and β, so are N'_2 and T_2, and hence, Eq. (7.7) determines X_2 as a function of ε and β. If the outside wires are to remain in contact with the center wire, $X_2 < 0$. When $X_2 > 0$; the outside wires separate from the center wire. Hence, setting $X_2 = 0$ in Eq. (7.7), yields ε as a linear function of β so that the outside wires are just touching the center wire. Figure 7.2 shows such a plot.

A consideration of the variations of the axial force F and the axial torque M_t for a strand leads to the equations of motion for a strand element. The resulting equations are

$$\frac{\partial F}{\partial \varepsilon}\frac{\partial \varepsilon}{\partial x} + \frac{\partial F}{\partial \beta}\frac{\partial \beta}{\partial x} = m\frac{\partial^2 u}{\partial t^2} \tag{7.8}$$

and

$$\frac{\partial M}{\partial \varepsilon}\frac{\partial \varepsilon}{\partial x} + \frac{\partial M}{\partial \beta}\frac{\partial \beta}{\partial x} = I\frac{\partial^2 \phi}{\partial t^2}, \tag{7.9}$$

which can be written as

$$a\frac{\partial^2 \bar{u}}{\partial \bar{x}^2} + b\frac{\partial^2 \phi}{\partial \bar{x}^2} = \frac{\partial^2 \bar{u}}{\partial \bar{t}^2} \tag{7.10}$$

and

$$c\frac{\partial^2 \bar{u}}{\partial \bar{x}^2} + d\frac{\partial^2 \phi}{\partial \bar{x}^2} = \frac{\partial^2 \phi}{\partial \bar{t}^2}, \tag{7.11}$$

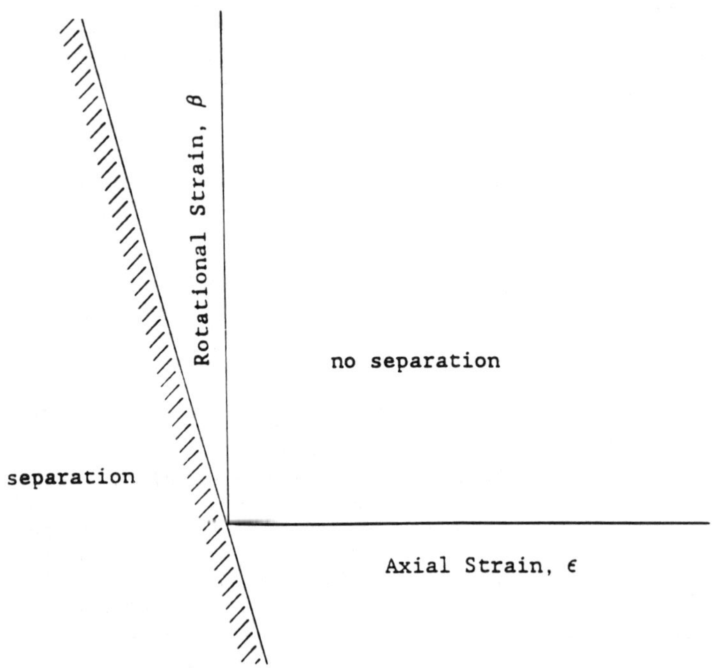

FIGURE 7.2. Plot showing regions of separation.

where

$$\bar{x} = \frac{x}{h}, \bar{R} = \frac{R}{h}, \bar{u} = \frac{u}{h}, \bar{F} = \frac{F}{AE}, \overline{M} = \frac{M}{ER^3}, \bar{t} = \frac{t}{\sqrt{\frac{mh^2}{AE}}}, \tag{7.12}$$

$$a = \frac{\partial \bar{F}}{\partial \varepsilon}, b = \bar{R}\frac{\partial \bar{F}}{\partial \beta}, c = \frac{mR^2h}{IA}\frac{\partial \overline{M}}{\partial \varepsilon}, d = \frac{mR^3h}{IA}\bar{R}\frac{\partial \overline{M}}{\partial \beta}, \tag{7.13}$$

x is the axial coordinate along the strand, $u = u(x, t)$ is the axial displacement, $\phi = \phi(x, t)$ is the axial rotation, t is time, A is the metallic area of the strand, m is the mass per unit length of the strand, and I is the mass moment of inertia per unit length of the strand. Since

$$\bar{F} = C_1\varepsilon + C_2\beta \tag{7.14}$$

and

$$\overline{M} = C_3\varepsilon + C_4\beta, \tag{7.15}$$

where the constants C_1, C_2, C_3, and C_4 are determined in Section 3.4, Eqs. (7.10) and (7.11) are a set of coupled linear partial differential equations for which the solution can be found for various boundary and initial conditions.

7.2 Solution of Equations

Solutions of the equations of motion [Eqs. (7.10) and (7.11)] with various initial and boundary conditions can be determined by Laplace transform techniques.

Consider, for example, a strand of length h, initially under a uniform tensile load and at rest so that

$$u(x, 0) = x\frac{\Delta}{h}, \quad \phi(x, 0) = 0, \quad \frac{\partial u}{\partial t}(x, 0) = 0, \quad \text{and} \quad \frac{\partial \phi}{\partial t}(x, 0) = 0, \tag{7.16}$$

where Δ is the axial deflection of the strand at $x = h$. Let the boundary conditions be given by

$$u(0, t) = 0, \quad \phi(0, t) = 0, \quad u(h, t) = f(t), \quad \text{and} \quad \phi(h, t) = 0, \tag{7.17}$$

where the form for $f(t)$ will be assumed later. In terms of the dimensionless variables, Eqs. (7.16) and (7.17) become

$$\bar{u}(\bar{x}, 0) = \bar{x}\bar{\Delta}, \quad \phi(\bar{x}, 0) = 0, \quad \frac{\partial \bar{u}}{\partial \bar{t}}(\bar{x}, 0) = 0, \quad \frac{\partial \phi}{\partial \bar{t}}(\bar{x}, 0) = 0, \tag{7.18}$$

and

$$\bar{u}(0, \bar{t}) = 0, \quad \phi(0, \bar{t}) = 0, \quad \bar{u}(1, \bar{t}) = \bar{f}(\bar{t}), \quad \phi(1, \bar{t}) = 0, \tag{7.19}$$

where $\bar{\Delta} = \Delta/h$ and $\bar{f} = f/h$.

Equations (7.10) and (7.11) become, after taking the Laplace transform,

$$a\frac{\partial^2 \tilde{u}}{\partial \bar{x}^2} + b\frac{\partial^2 \tilde{\phi}}{\partial \bar{x}^2} = s^2\tilde{u} - s\bar{x}\bar{\Delta} \tag{7.20}$$

and

$$c\frac{\partial^2 \tilde{u}}{\partial \bar{x}^2} + d\frac{\partial^2 \tilde{\phi}}{\partial \bar{x}^2} = a^2\tilde{\phi}, \tag{7.21}$$

where

$$\tilde{u}(\bar{x}, s) = L(\bar{u}) = \int_0^\infty \bar{u}(\bar{x}, \bar{t})e^{-s\bar{t}}\, d\bar{t}. \tag{7.22}$$

The homogeneous solution of Eqs. (7.20) and (7.21) is obtained by assuming a solution of the form

$$\tilde{u} = C_k e^{r_k \bar{x}} \quad \text{and} \quad \tilde{\phi} = d_k e^{r_k \bar{x}}. \tag{7.23}$$

A substitution of Eq. (7.23) into Eqs. (7.20) and (7.21) (the homogeneous equations) yields

$$r_k^2 = \frac{(a+d)s^2 \pm s^2\sqrt{(a+d)^2 - 4(ad-bc)}}{2(ad-bc)} \tag{7.24}$$

and

$$d_k = \frac{(s^2 - ar_k^2)}{br_k^2} c_k. \tag{7.25}$$

The particular solution is

$$\tilde{u}_p = \frac{\bar{x}\bar{\Delta}}{s} \quad \text{and} \quad \tilde{\phi}_p = 0. \tag{7.26}$$

Equation (7.24) yields four roots and, hence, the solution for the transformed variables can be written as

$$\tilde{u}(\bar{x}, s) = c_1(s)e^{e_1 s\bar{x}} + c_2(s)e^{-e_2 s\bar{x}} + c_3(s)e^{e_2 s\bar{x}} + c_4(s)e^{-e_2 s\bar{x}} + \bar{x}\frac{\bar{\Delta}}{s} \tag{7.27}$$

and

$$\tilde{\phi}(\bar{x}, s) = d_1(s)e^{e_1 s\bar{x}} + d_2(s)e^{e_1 s\bar{x}} + d_3(s)e^{e_2 s\bar{x}} + d_4(s)e^{-e_2 s\bar{x}}, \tag{7.28}$$

where

$$e_1 = \sqrt{\frac{(a+d) + \sqrt{(a+d)^2 - 4(ad-bc)}}{2(ad-bc)}} \tag{7.29}$$

and

$$e_2 = \sqrt{\frac{(a+d) + \sqrt{(a+d)^2 - 4(ad-bc)}}{2(ad-bc)}}. \tag{7.30}$$

Equation (7.25) yields

$$d_1 = g_1 c_1, \quad d_2 = g_1 c_2, \quad d_3 = g_2 c_3, \quad \text{and} \quad d_4 = g_2 c_4, \tag{7.31}$$

where

$$g_1 = \frac{1 - ae_1^2}{be_1^2} \quad \text{and} \quad g_2 = \frac{1 - ae_2^2}{be_2^2}. \tag{7.32}$$

A satisfaction of the transformed boundary conditions

$$\tilde{u}(0,s) = 0, \quad \tilde{\phi}(0,s) = 0, \quad \tilde{u}(1,s) = \tilde{f}(s), \quad \text{and} \quad \tilde{\phi}(1,s) = 0 \tag{7.33}$$

results in

$$c_1(s) = \frac{g_2}{g_2 - g_1}\left[\tilde{f}(s) - \frac{\bar{\Delta}}{s}\right]\frac{1}{e^{e_1 s} - e^{-e_1 s}},$$

$$c_2(s) = -\frac{g_2}{g_2 - g_1}\left[\tilde{f}(s) - \frac{\bar{\Delta}}{s}\right]\frac{1}{e^{e_1 s} - e^{-e_1 s}},$$

$$c_3(s) = \frac{g_1}{g_1 - g_2}\left[\tilde{f}(s) - \frac{\bar{\Delta}}{s}\right]\frac{1}{e^{e_2 s} - e^{-e_2 s}},$$

and

$$c_4(s) = -\frac{g_1}{g_1 - g_2}\left[\tilde{f}(s) - \frac{\bar{\Delta}}{s}\right]\frac{1}{e^{e_3 s} - e^{-e_2 s}}, \tag{7.34}$$

where the exponential factors appearing in Eq. (7.34) can be expanded in a power series; that is,

$$\frac{1}{e^{e_k s} - e^{-e_k s}} = \sum_{n=0}^{\infty} e^{-(2n+1)e_k s}. \tag{7.35}$$

Since the solution for $\tilde{u}(\bar{x}, s)$ and $\tilde{\phi}(\bar{x}, s)$, is known, the inverse transform can be obtained. The result is

$$\begin{aligned}
&\bar{u}(\bar{x}, \bar{t}) \\
&= \frac{g_2}{g_2 - g_1}\sum_{n=0}^{\infty}\{\bar{f}[\bar{t} - (2n+1-\bar{x})e_1] - \bar{\Delta}\}H[\bar{t} - (2n+1-\bar{x})e_1] \\
&\quad - \frac{g_2}{g_2 - g_1}\sum_{n=0}^{\infty}\{\bar{f}[\bar{t} - (2n+1+\bar{x})e_1] - \bar{\Delta}\}H[\bar{t} - (2n+1+\bar{x})e_1] \\
&\quad + \frac{g_1}{g_1 - g_2}\sum_{n=0}^{\infty}\{\bar{f}[\bar{t} - (2n+1-\bar{x})e_2] - \bar{\Delta}\}H[\bar{t} - (2n+1-\bar{x})e_2] \\
&\quad - \frac{g_1}{g_1 - g_2}\sum_{n=0}^{\infty}\{\bar{f}[\bar{t} - (2n+1+\bar{x})e_2] - \bar{\Delta}\}H[\bar{t} - (2n+1+\bar{x})e_1] \\
&\quad + \bar{x}\bar{\Delta}
\end{aligned} \tag{7.36}$$

and

$$
\begin{aligned}
\phi(\bar{x},\bar{t}) \\
&= \frac{g_1 g_2}{g_2 - g_1} \sum_{n=0}^{\infty} \{\bar{f}[\bar{t} - (2n + 1 - \bar{x})e_1] - \bar{\Delta}\} H[\bar{t} - (2n + 1 - \bar{x})e_1] \\
&\quad - \frac{g_1 g_2}{g_2 - g_1} \sum_{n=0}^{\infty} \{\bar{f}[\bar{t} - (2n + 1 + \bar{x})e_1] - \bar{\Delta}\} H[\bar{t} - (2n + 1 + \bar{x})e_1] \\
&\quad + \frac{g_1 g_2}{g_1 - g_2} \sum_{n=0}^{\infty} \{\bar{f}[\bar{t} - (2n + 1 - \bar{x})e_2] - \bar{\Delta}\} H[\bar{t} - (2n + 1 - \bar{x})e_2] \\
&\quad - \frac{g_1 g_2}{g_1 - g_2} \sum_{n=0}^{\infty} \{\bar{f}[\bar{t} - (2n + 1 + \bar{x})e_2] - \bar{\Delta}\} H[\bar{t} - (2n + 1 + \bar{x})e_2],
\end{aligned}
\tag{7.37}
$$

where $H(t)$ is the unit step function.

The local strains $\varepsilon = \partial\bar{u}/\partial\bar{x}$ and $\beta = \partial\phi/\partial\bar{x}$ can be obtained by differentiating Eqs. (7.36) and (7.37).

7.3 Numerical Results

The following numerical results are presented for a given steel strand with the following characteristics:

$$
\begin{aligned}
h &= 120 \text{ in.} \\
R_1 &= 0.03155 \text{ in.} \\
R_2 &= 0.02890 \text{ in.} \\
p_2 &= 1.30 \text{ in.} \\
\nu &= 0.25 \\
E &= 30 \times 10^6 \text{ psi} \\
m_2 &= 6
\end{aligned}
$$

Let $\Delta = 0.360$ in. and

$$
f(t) = \begin{cases} \left(1 - \dfrac{t}{2\bar{T}}\right)\Delta & \text{for} \quad 0 \le t \le \bar{T}, \\[2ex] \dfrac{\Delta}{2} & \text{for} \quad t \ge \bar{T}. \end{cases}
$$

Also, let the ends of the strand be prevented from rotation, let the strand be initially at rest under a uniform tension corresponding to the displacement Δ at the end $x = h$, and let the end $x = 0$ be fixed. This corresponds to a strand initially under a uniform tensile load and twisting moment. The initial strains are $\varepsilon = 0.003$ and $\beta = 0$. The end at $x = h$ is displaced in a linear manner so as to reduce the tensile load. After a time, $\bar{T}$, the end is held fixed at the displacement $\Delta/2$. The quantity $\bar{T}$ then determines the velocity of the impacted end of the strand. The equation for the straight line shown in Fig. 7.2 for the given strand is $\beta = -4.6805\varepsilon$. If the impacted end has a velocity of 23.47 fps, the strains at the end $x = 0$, after the first reflection, are $\varepsilon = 0.000257$ and $\beta = -0.00102$, which indicate a point in Figure 7.2, where the wires do not separate. If the impacted velocity is increased to 24.2 fps, the theory indicates a strain of $\varepsilon = 0.000172$, and $\beta = -0.00105$, which is in the separation zone. An impacted velocity of about 23.88 fps yields $X_2 = 0$ at the fixed end $x = 0$, with a strain of $\varepsilon = 0.0002232$.

The above theory indicates wire separation even if the axial strain is positive. The wires can separate, due to twisting, with a reduction in the tensile load. If a rope is carrying a heavy load and the load is suddenly partially released, wire separation is possible.

The above theory can be extended to wire rope with complex cross sections. The form of the equations remains the same. It is possible, however, to have the wires or strands separate and still not form a bird cage, for, as noted previously, a bird cage is a permanent condition and thus would require inelastic behavior in the wires.

8 Rope Rotation

8.1 Rotation of a Wire Rope

As was shown previously, the total axial force F and the total axial twisting moment M_t acting on a rope can be expressed as

$$\frac{F}{AE} = C_1\varepsilon + C_2\beta \tag{8.1}$$

and

$$\frac{M_t}{ER^3} = C_3\varepsilon + C_4\beta, \tag{8.2}$$

where $A = \Sigma\pi R_i^2$, R is the radius of the rope and ε and β are the axial and rotational strains. The rotational strain is defined by the equation

$$\beta = R\tau, \tag{8.3}$$

where τ is the angle of twist per unit length.

The above equations enable one to determine how much one cross section rotates relative to another under a given loading system. For example, in cord composites, the rotation of cord becomes important in determining the axial stiffness of the cord. Also, some ropes are joined together by making a loop or an eye in the ends of the ropes and then tucking the ends of the strands back into the main body of the rope (a splice). If one end of the rope rotates with respect to the other end, the efficiency of such a splice can be significantly reduced [25].

Example 8.1. Consider a simple straight strand with $R_1 = 0.103$ in., $R_2 = 0.101$ in., $p_2 = 9.75$ in., $E = 28{,}500{,}000$ psi, $\nu = 0.25$, and $m_2 = 6$. This is the strand considered in Example 3.3. The results are

$$\frac{F}{AE} = 0.975\varepsilon + 0.0728\beta \tag{a}$$

and

$$\frac{M_t}{ER^3} = 0.167\varepsilon + 0.0694\beta. \tag{b}$$

Now let $M_t = 0$. Equation (b), above, yields $\beta = -2.406\varepsilon$ and, hence, Eq. (a), above, becomes

$$\frac{F}{AE} = 0.800\varepsilon. \tag{c}$$

If $F = 10{,}000$ lb, $\varepsilon = 0.00194$, $\beta = -0.00468$, and the angle of twist per unit length becomes

$$\frac{\theta}{h} = -0.0153 \text{ rad/in.} \tag{d}$$

If the strand, for example, were 30 in. long, the rotation of one end of the strand with respect to the other would be

$$-0.0153 \times 30 \times \frac{360}{2\pi} = -26.35 \text{ degrees.} \tag{e}$$

When a rope is allowed to rotate fully, the results are even more pronounced. This is because a rope has relatively finer wires in a given cross section than does a simple strand. For example, the equations for the 6 × 19 Seale IWRC wire rope considered in Chapter 4 are

$$\frac{F}{AE} = 0.7020\varepsilon + 0.1232\beta \tag{f}$$

and

$$\frac{M_t}{ER^3} = 0.2060\varepsilon + 0.0403\beta, \tag{g}$$

and, hence, for $M_t = 0$,

$$\frac{F}{AE} - 0.0722\varepsilon, \tag{h}$$

which indicates a large reduction in the effective modulus. Also $\beta = -5.112\varepsilon$.

Example 8.2. Consider a right-lay 6 × 19 Seale IWRC wire rope is 30 ft long and 3 in. in diameter. Let this rope be attached to a left-lay 6 × 19 Seale IWRC wire rope, 40 ft long and 2 in. in diameter. Let the rope be loaded so that both ends of the 70-ft rope are prevented from rotating.

The equations for the right-lay rope are

$$\frac{F}{AE} = 0.7020\varepsilon + 0.1232\beta \tag{a}$$

and

$$\frac{M_t}{ER^3} = 0.2060\varepsilon + 0.0403\beta; \tag{b}$$

for the left-lay rope,

$$\frac{F}{AE} = 0.7020\varepsilon - 0.1232\beta \tag{c}$$

and

$$\frac{M_t}{ER^3} = -0.2060\varepsilon + 0.0403\beta. \tag{d}$$

The sign convention for positive ε, β, F, and M_t are the same as that used in Chapter 3.

Let the 70-ft rope be loaded with a 250,000-lb load and left the metallic area of the right-lay rope be $A_r = 4.19$ in.^{2} and the left-lay rope, $A_l = 1.86$ in.^{2}. Equations (a), (b), (c), and (d) become

$$\frac{250{,}000}{4.19 \times 30 \times 10^6} = 0.7020\varepsilon_r + 0.1232 \times \frac{3.00}{2}\frac{\theta}{30 \times 12},$$

$$\frac{M_t}{30 \times 10^6 \times \left(\frac{3}{2}\right)^3} = 0.2060\varepsilon_r + 0.0403 \times \frac{3.00}{2}\frac{\theta}{30 \times 12},$$

$$\frac{250{,}000}{1.86 \times 30 \times 10^6} = 0.7020\varepsilon_l + 0.1232 \times \frac{2.00}{2}\frac{\theta}{40 \times 12},$$

and

$$\frac{M_t}{30 \times 10^6 \times \left(\frac{2}{2}\right)^3} = 0.2060\varepsilon_l - 0.0403 \times \frac{2.00}{2}\frac{\theta}{40 \times 12},$$

where it is assumed that at the connection both ends of the rope rotate through the same angle θ. The solution to the above equations for θ is

$$\theta = -3.0085 \text{ rad} = -172.37 \text{ degrees},$$

which means that both sections of the rope unwind. Also $M_t = 53{,}823$ in. lb, $\varepsilon_r = 0.00503$, and $\varepsilon_l = 0.00748$. Both sections of the rope should be checked to see if the contact force between the strands and IWRC is negative.

8.2 Hand-Spliced Ropes

In an article entitled "Warning: Hand-Spliced Slings and Rotating Loads—a Bad Combination" by Gail Dull and Robin Parkinson [25], the author hammer home an inviolable commandment: Thou shalt not use single-lift

TABLE 8.1. Sling efficiencies for various test procedures.

Sling identification	Straight pull	One end free to rotate	Load-unload cycle with rotation	Rope breaking strength (lb)
Standard Eye	71.97	35.84	12.83	45,200
Cargo-Marine	73.98	73.61	56.69	45,200
Admirality	76.88	68.92	67.92	45,200
Wallace Hidden Tuck	60.27	49.80	44.16	45,200
Loggers	67.39	68.45	56.11	45,200
Lock-tuck	94.39	59.91	49.36	43,700*
Tapered and Concealed	77.40	49.80	68.68	44,700*
Liverpool	89.45	58.05	50.27	43,700*
Australian	—	—	49.77	43,700*
Press-Grip**	91.65	83.43	74.56	45,200

* Estimated rope breaking strengths.
** Mechanically spliced sling.

hand-spliced slings where the sling may rotate." In this article, tests were run on ropes to determine the sling efficiencies of various hand-spliced ropes. Table 8.1, taken from the above-mentioned article, shows the efficiencies for various types of hand-spliced slings.

The results show that a rather large drop in the efficiencies can occur if one end of the rope is free to rotate relative to the other. As the rope rotates, one end relative to the other, the strands tend to straighten out. This will reduce the line contact load between the strand and the core. Since a splice depends upon friction to sustain the load, this reduction in contact load will reduce the allowable friction load and hence the axial load will be reduced. Sometimes the unwinding is severe and, hence, the large strain theory should be used.

References

1. *Wire Rope Handbook*, St. Joseph, Mo.: Leschen Wire Rope Company, 1971.
2. Scalzi, J.B. and McGrath, W.K. Mechanical properties of structural cables, *Journal of the Structural Division, ASCE*, 97, 2837–2844, 1971.
3. Sayenga, D. The Birth and Evaluation of the American Wire Rope Industry, *First Annual Wire Rope Proceedings*, Engineering Extension Service, Washington State University, Pullman, Wash. 99164, March 1980.
4. Wire Rope Users Manual (Washington, D.C.: American Iron and Steel Institute, 1979.
5. Love, A.E.H. *A Treatise on the Mathematical Theory of Elasticity*, New York: Dover Publications, 1944, Chaps. 18 and 19.
6. Boresi, A.P. and Sidebottom, O.M. *Advanced Mechanics of Materials*, New York: John Wiley and Sons, 1985, Chap. 14.
7. Eisenhart, L.P. *An Introduction to Differential Geometry*, Princeton, N.J.: Princeton University Press, 1940, pp. 25–27.
8. Costello, G.A. Large deflections of helical spring due to bending, *Journal of the Engineering Mechanics Division, ASCE*, 103, (EM3, Proc. Paper 12964), 479–487, 1977.
9. Costello, G.A. and Butson, G.J. A simplified bending theory for wire rope, *Journal of the Engineering Mechanics Division, ASCE*, 108, (EM2, Proc. Paper 16984), 219–227, 1982.
10. McConnell, K.G. and Zemke, W.P. The measurement of flexural stiffness of multi-stranded electrical conductors while under tension, Experimental Mechanics, 20(6), 198–204, 1980.
11. Costello, G.A. Stresses in multilayered cables, *Journal of Energy Resources Technology, Trans. the ASME*, 105, 337–340, 1983.
12. Costello, G.A. and Phillips, J.W. A more exact theory for twisted wire cables, *Journal of the Engineering Mechanics Division, ASCE*, 100 (No. EM5, Proc. Paper 10856), 1096–1099, 1974.
13. Costello, G.A. and Miller, R.E. Lay effect of wire rope, *Journal of the Engineering Mechanics Division, ASCE*, 105 (No. EM4, Paper 14753), 597–608, 1979.
14. Velinsky, S.A. Analysis of wire ropes with complex cross sections," Ph.D. thesis, Department of Theoretical and Applied Mechanics, University of Illinois at Urbana-Champaign, 1981, 87.
15. Costello, G.A. and Phillips, J.W. Stress analysis of wire hoist rope, Technical Report No. UILU-ENG 83-6006, Engineering Documents Center, University of Illinois at Urbana-Champaign, Urbana, Ill. Sept. 1983, 103.

16. LeClair, R.A. and Costello, G.A. Axial bending and torsional loading of a strand with friction, *Proceedings of the Fifth International OMAE Symposium, ASME*, Vol. III, pp. 550–555, Tokyo Japan, 1986.
17. Hobbs, R.E. and Ghavami, K. The fatigue of structural wire strands, *International Journal of Fatigue*, 4, (2), 69–72, 1982.
18. Chien, C.H. and Costello, G.A. Effective length of a fractured wire in wire rope, *Journal of the Engineering Mechanics Division, ASCE*, III(7) 952–961, 1985.
19. Langhaar, H.L. *Dimensional Analysis and Theory of Models*, Huntington, N.Y., Robert E. Krieger Publishing Company, 1980, 166 p.
20. *Yellow Strand Wire Rope Handbook*, Broderick and Bascom Rope Company, St. Louis, Miss.
21. Chien, C.H. Effective length of fractured wires and a fatigue analysis of wire rope, Ph.D. thesis, Department of Theoretical and Applied Mechanics, University of Illinois at Urbana-Champaign, 1984, 104 p.
22. Beeman, G.H. Factors affecting the service life of large-diameter wire rope, Report to the U.S. Department of Energy by Pacific Northwest Laboratory, Contract No.-ET-75-C-01-9099, March 1978.
23. Morganstern, M.H., et al. Wire rope improvement program, fiscal years 1979–1980, "Interim Report to the U.S. Department of Energy by Pacific Northwest Laboratory, Contract No. DE-A106-76RLD 1830, Aug. 1980.
24. Drucker, D.C. and Tachau, H. A new design criterion for wire rope, *Journal of Applied Mechanics, Trans. ASME*, 67, A-337, A-38, 1945.
25. Dull, G. and Parkinson, R Warning: Hand-spliced slings and rotating loads—a bad combination, *Wire Rope News and Sling Technology*, April 1983, 18–22.

Additional References

Albert, W.A.J. On the manufacture of whim ropes from iron wire, *The Mining Journal and Commerical Gazette, Suppl. XII*, Feb. 25, 1837 (extracts from Foreign Scientific Works V), pp. 47–48.

Scoble, W.A. First report of wire rope research committee, *Proceedings of the Institution of Mechanical Engineers*, 115, 835–868, 1920; second report, 119, 1193–1290, 1924; third report, 123, 353–404, 1928; fourth report, 125, 553–602, 1930; fifth report, 130, 373–478, 1935.

Skillman, E. Some tests of steel wire rope on sheaves, U.S. Bureau of Standards, Technologic Paper, No. 229, March 2, 1923, pp. 227–243.

Woernle, R. Drahlseilforschung, *Zeitschrift des Vereines Deutscher Ingenieure*, 73, 417–426, 192; 73, 1623–1624, 1929; 74, 185–186, 1930; 74, 1417–1419, 1930; 206–209, 1931, 75, 1485–1489, 1931; 76, 557–560, 1932; 77, 799–803, 1933; 78, 1492–1489, 1934; 79, 1281–1282, 1935; 83, 1056, 1939.

de Forest, A.V. and Hopkins, L.W. Testing of rope wire and wire rope, *Proceedings of ASTM* 32, 398–412, 1932.

Suslov, B.M. On the modulus of elasticity of wire ropes, *Wire and Wire Products*, 11, 176–182, 1936.

Hansom, O.P. Mechanics of locked coil steel wire ropes, Ph.D. thesis, University of Birmingham, 1948.

Matheson, J.A.L. The mechanics of locked coil ropes, *Engineering*, June 18 and 25, 578–581 and 601–604, 1948.

Slight, G.C. The torsional properties of three and seven strand wire ropes with a view to their use in multiple strand helical springs, M.Sc. Thesis, University of London (Woolwich Polytechnic), 1949.

Hall, H.M. Stresses in small wire ropes, *Wire and Wire Products*, 26, 766–767, 799–801, 1951.

Forestier-Walker, E.R. *A History of the Wire Rope Industry in Great Britain*, British Wire Rope Manufacturers, 1952.

Hruska, F.H. Radial forces in wire ropes, *Wire and Wire Products*, 27, 459–463, 1952.

Hruska, F.H. Tangential forces in wire ropes, *Wire and Wire Products*, 28, 455–460, 1953.

Boyer, W.A. Safety factor characteristic curves: Their application to mine hoisting ropes, *Mixing Engineering, Trans. AIME*, Oct. 1954, 989–993.

Boyer, W.A. Safety factor characteristic curves for mine hoisting ropes, *Mining Engineering, Trans. AIME*, March 1956, pp. 307–309.

Krolevets, M.S. The modulus of elasticity in steel wire ropes, Issled po voprosam ustoichovosti i prochnosti, Kiev, AN, USSR, 1956, pp. 243–253 (SMRE, Health and Safety Exec trans. No. 4269).

Leissa, A.W. Contact stresses in wire ropes, *Wire and Wire Products*, 34, 307–314, 372–373, 1959.

Starkey, W.L. and Cress, H.A. An analysis of critical stresses and mode of failure of a wire rope, *Journal of Engineering for Industry, Trans. ASME* 81, 307–316, 1959.

Nesterov, P.O., Shabanov-Kushnarenko, Yu.P. and Kozyuberda, N.I. A new method of determination of stresses in wire ropes, trans. from the Russian, *Zavodskaya Laboratoriya*, 27, (2) 191–194, 1961.

Glushko, M.F. Mechanical testing of wire ropes, trans. from the Russian, *Zavodskaya Laboratoriya*, 28 (8), 981–983, 1962.

Bert, C.W. and Stein, R.A. Stress analysis of wire rope in tension and torsion, *Wire and Wire Products*, 37, (5), 621–624, (6), 769–770, 772–816, 1962.

Bechtloff, G. Longitudinal elongation and transverse contraction of a six-stranded wire rope under tensile load, *Wire World*, 5, (6), 1963.

Dong, R.G. and Steidel, R.F. Contact Stress in stranded cable, *Experimental Mechanics*, 5, (5), 142–147, 1965.

Martin, B.C. and Packard, T.J. Stresses in wire strand, BSc Project Report, University of Bristol, Department of Civil Engineering, England, 1966.

Gambrell, S.C., Jr. and Case, R.O. New machine for accelerated fatigue tests of wire rope, *Wire and Wire Products*, June 1968, pp. 46–49.

Lutchansky, M. Axial stresses in armor wires of bent submarine cables, *Journal of Engineering for Industry, Trans. ASME*, Aug. 1969, 688–691.

Gambrell, S.C., Jr. Study low-cycle fatigue of wire rope, *Wire and Wire Products*, Oct. 1969, pp. 127–130.

DeRuntz, J.A., Jr. End Effect Bending Stresses in Cables, *Journal of Applied Mechanics, Trans. ASME*, Dece. 1969, pp. 750–756.

Gibson, P.T., Cress, H.A., Kaufmann, W.J., and Gallant, W.E. Torsional properties of wire rope, ASME Paper No. 69-DE-34, *Proc. Design Eng. Div. Conference*, New York, 1969; Analysis of wire rope torque, *Wire and Wire Products*, 45(11), 50, 52–58, 60, 1970.

Laura, P.A. Vanderveidt, H., and Gaffney, P. Mechanical behaviour of stranded wire rope, *Marine Tech Soc Journal*, 4(3) 19–32, 1970.

Gambrell, S.C., Jr. Predicting fatigue life of wire rope from tests on single wire, *Wire and Wire Products*, Nov. 1970, pp. 45–49.

Chi, M. Analysis of operating characteristics of strands in tension allowing end rotation, ASME Paper No. 73-WA/OCT-19, *Proc. ASME Oc Tech Div Meeting*, New York, 1972.

Durelli, A.J. Machida, S., and Parks, V.J. Strains and displacements on a steel wire strand, *Naval Engineers Journal*, 84, (6), 85–93, 1972.

Christen, R. and Oplatka, G. *Tail cones*: Does doubling back of the wire increase safety, *Wire* 23, (4) 160–164, 1973.

Vanderveldt, H.H., Chung, B.S., and Reader, W.T. Some dynamic properties of axially loaded wire ropes, *Experimental Mechanics*, Jan. 1973, pp. 24–30.

Costello, G.A., and Phillips, J.W. Contact stresses in thin twisted rods, *Journal of Applied Mechanics, Trans. American Society of Mechanical Engineers* (Series E.) 40 629–630, 1973.

Durelli, A.J. and Machida, S. Response of epoxy oversized models of strands to axial and torsional loads, *Experimental Mechanics*, 13, 313–321, 1973.

Hankus, J. Spinning moment in winding ropes, (in Polish with English abstract), Glowny Instytut Gornictwa, Katowice 1973, Kommunikat, No. 579.

Hilgers, W. Wire rope tail cones and alloys, *Wire*, 24, (6), 251–263, 1973.

Machida, S. and Durelli, A.J. Response of a strand to axial and torsional displacements, *Journal of Mechanical Engineering Science*, 15(4), 241–251, 1973.

Mancini, G. and Rossetti, U. Sur l'analyse es constraintes et does deformations des cables flechis, *Proceedings OIPEEC Round Table*, Milan 1973, pp. 65–86.

Paolini, G. and Bazzarao, E. Study on the state of stress in the wires of steel ropes under tensile loads, *Proceedings OIPEEC Round Table*, Milan, 1973, pp. 112–122.

Phillips, J.W. and Costello, G.A., Contact stresses in twisted wire cables, *Proceedings ASCE, Journal Eng Mech Div*, 99(No. EM2), 31–341. 1973.

Chi, M. Analysis of multi-wire strands in tension and combined tensionand torsion, *Proceedings of the Seventh South Eastern Conference on Theoretical and Applied Mechanics*, 7, 599–639, 1974.

Laura, P.A. Un Resumen de Recientes Investigaciones Analitieas Y Experimentales Sobre Cables Ocenogra fic os, *De Anales De La Sociedad Cientifica Argentina*. T/CXCVIII, Entrega IV-VI, 1974, pp. 67–86.

Nowak, G. Computer design of electromechanical cables for ocean application, *Proceedings of 10th Annual Conference, Marine Tech Society*, Washington, D.C., 1974, pp. 293–305.

Samras, R.K., Skop, R.A., and Milburn, D.J.A. An analysis of coupled extensional-torsional oscillations in wire rope, *Journal of Engineering for Industry, Trans. American Society of Mechanical Engineers*, 96, 1130–1135, 1974; AMR 28, (1975), Rev. 8795.

Weber, W. Development and production of wire rope I and II, *Wire World International* (I), 16, 286–291, 1974; (II), 17, 20–24, 1975.

Shelley, P.D. An analysis of tail rope behaviour, *The South African Mechanical Engineer*, 24, 310–327, 1974.

Wiek, L. Facts and figures of stresses in ropes, I and II, *Wire* (I), 26(4), 173–178, 1975; (II), 26(5), 214–216, 1975.

Wiek, L. Measured differences between steel wire ropes in normal lay and in lang lay, Report Technische Hogeschool, Delft, Transportkunde, Oct. 1975.

Knapp, R.H. Non-linear analysis of a helically armoured cable with non-uniform mechanical properties in tension and torsion, IEEE paper No. 75CHO, 995-1, OEC, *Proceedings IEEE Conference on Engineering in the Ocean, Environmental and Marine*, Tech Sec. 11th Annual Meeting, San Diego, California, 1975, pp. 155–164.

Costello, G.A. and Phillips, J.W. Effective modulus of twisted wire cables, *Proceedings ASCE, Journal Engineering Mechanical Division*, 102(No. EM1), 171–181, 1976.

Kollros, W. Relationship between torque, tensile force and twist in wire ropes, *Wire*, 26(1), 19–24, 1976.

Costello, G.A. and Sinha, S.K. Torisonal stiffness of twisted wire cables, *Proceedings ASCE, Journal Engineering Mechanical Division*, 103(No. EM4), 766–770, 1977.

Costello, G.A. and Sinha, S.K. Static behaviour of wire ropes, *Proceedings ASCE, Journal Engineering Mechanical Division*, 103(No. EM6), 1011–1022, 1977.

Hankus, J., The permanent and percentage elongation of winding ropes in factory condition, translated from the Polish, Glowny Instytut Gornictwa Katowice, 1977, Kommunikat No. 682, pp. 3–14, (SMRE, Health and Safety Exec translation No. MRDE 1054).

Matanzo, F., Jr. and Metcalf, J.T., Jr. Efficiency of wire rope terminations used in the mining industry, *Proceedings OIPEEC Round Table*, Luxemburg, 1977. Also *Journal of Engineering Materials and Technology, Trans. ASME*, 103, 164–170, 1981.

Myers, W.H. Major part of lifting devices; Wire rope end attachments, *Wire Journal*, 10(No. 3), 67–71, 1977.

Phillips, J.W. and Costello, G.A. Axial impact of twisted wire cables, *Journal of Applied Mechanics, Trans. American Society of Mechanical Engineers*, 44, 127–131, 1977.

Treloar, L.R.G. Physics of textiles, *Physics Today*, 30, 23–30, 1977.

Wiek, L. The influence of broken wires on wire rope strength and discarding, *Proceedings OIPEEC Round Table*, Luxemburg, 1977, Paper No. 4-4.

Costello, G.A. Analytical investigation of wire rope, *Applied Mechanics Reviews*, 31(7), 897–900, 1978.

Hankus, J. Examination of elastic modulus of mine winding ropes in conditions of dynamic loading (in Polish with English abstract) Katowice 1978, Glowny Instytut Gornictwa, Kommunikat No. 700.

Hankus, J. Elastic modulus of mine winding ropes in conditions of static loading, (in Polish with English abstract), Glowny Instytut Gornictwa, Katowice 1978, Kommunikat No. 695.

Huang, N.C. Finite extension of an elastic strand with a central core, *ASME Journal of Applied Mechanics*, 45(4), 852–858, 1978.

Babel, H. Destructive and non-destructive test methods to determine the life of wire ropes I and II, *Wire* (I), 28(6), 263–270, 1979; (II) 29(1) 38–44, 1980.

Gathman, D.W. Resin socketing for wire rope attachments, *Wire Journal*, 12(6), 82–85, 1979.

Knapp, R.H. Derivation of a new stiffness matrix for helically armoured cables considering tension and torsion, *International Journal for Numerical Methods in Engineering*, 14, 515–529, 1979.

Sharp, D.M. Wire rope in the marine environment I and II, *Wire Industry* (I), 46(543-1979), 198–202, 1979; (II) 46(544), 270–272, 1979.

Stonesifer, F.R. and Smith, H.L. Tensile fatigue in wire rope, *Proceedings Offshore Technology Conference*, 1, 539–545, 1979.

Wiek, L. Strain gauge measurements at multi-strand non-spinning ropes, Publication No. 212, Technische Hogeschool, Delft, Transportkunde, 1979.

Bahke, E. Principles defining the strength of wire ropes and chains I and II, *Wire* (I) 29(2), 54–61, 1980; (II) 30(3), 168–176, 1980.

Costello, G.A. and Miller, R.E. Static response of reduced rotation rope, *Proceedings ASCE, Journal Engineering Mechanics Division*, 106(No. EM4), 623–631, 1980.

Gibson, P.T. Wire rope behaviour in tension and bending, *Proceedings of the First Annual Wire Rope Symposium, Denver, Colorado.* Published by Engineering Extension Service, Washington State University, Pullman, Wash., March 1980, pp. 3–31.

Molinari, G. Experimental research on the strains distribution in the wires of a simple spiral strand under tensile loading during the breaking of one of them, (in Italian), *Elevatori*, 6, 30–39, 1980.

Phillips, J.W. Miller, R.E. and Costello, G.A. Contact stresses in straight cross lay wire rope, *Proceedings of the First Annual Wire Rope Symposium, Denver, Colorado.* Published by Engineering Extension Service, Washington State University, Pullman, Wash., March 1980, pp. 177, 199.

Velinsky, S. and Costello, G.A. Axial response of oval wire ropes, ASME Paper No. 80-WA/OCE-3. *Proceedings ASME Ocean Engineering Division Conference*, Chicago, Illinois, 1980.

Rice, R.C. The correlation of large and small diameter wire rope bending fatigue behavior, *Proceedings of the First Annual Wire Rope Symposium*, sponsored by Washington State University at Denver, Colorado, March 1980, pp. 48–71.

Karamehetty, D. Some geometrical characteristics of wires in wire ropes and cables, *Wire Journal*, Nov. 1980, pp. 98–104.

Boyle, J.A. Recent trends in the development of wire rope strands, *Wire Industry*, 48(565), 37–38, 1981.

Dodd, J.M. Resin as a socketing medium, *Wire Industry*, 48, 343–344, 1981.

Knapp, R.H. Torque and stress balanced design of helically armoured cables, *Journal of Engineering for Industry, Trans ASME*, 103(1), 61–66, 1981.

Matanzo, F., Jr. and Metcalf, J.T., Jr. Efficiency of wire rope terminations used in the mining industry, *Proceedings OIPEEC Round Table*, Luxemburg, 1977. Also *Journal of Engineering Materials and Technoogy, Trans ASME*, 103, 164–170, 1981.

Wiek, L. Stress deviations in steel wire ropes, *Proceedings OIPEEC Round Table*, Cracow, 1981.

Hobbs, R.E. and Raoof, M. Interwire slippage and fatigue prediction in stranded cables for TLP tethers, in *Behaviour of Offshore Structures, Vol. 2*, Chryssostomiolis, C. and Connor, J.J. (Eds.), New York: Hemisphere Publishing/McGraw-Hill, 1982, pp. 77–99. (*Proceedings* 3rd Int Conf *on Behaviour of Offshore Structures*, 1982).

Hanzawa, M., Yokota, H., Toda, Y., and Yokoyama, K. Fatigue behavior of large-diameter ropes, *Society of Petroleum Engineers Journal*, 22k(3), 420–428, 1982.

McConnell, K.G. and Zemke, W.P. A model to predict the coupled axial torsion properties of ACSR electrical conductors, *Experimental Mechanics*, July 1982, pp. 237–244.

Velinsky, S.A., Phillips, J.W., Anderson, G.L., and Costello, G.A. Axial wire stresses in rope, *Proceedings 9th US National Congress of Applied Mechanics*, June 1982, Ithaca, New York, 3 p.

Chaplin, C.R. and Sharman, R.C. Mechanisms of load transfer in resin socketed terminations, *International Wire and Mach. Assoc. Conference on Offshore Applications*, Aberdeen, 1983.

Hobbs, R.E. and Smith, B.W. Fatigue performance of socketed terminations to structural strands, (Proc. Instn. Civ. Engrs., Part 2, March 1983, 75, pp. 35–48.

Hobbs, R.E. and Raoof, M. Prediction of elastic properties of large strands and cables, Int. Wire and Mach Assoc Conference on "Offshore Applications," Aberdeen, 1983, pp. 1.3.1–1.3.16.

Ropeman's Handbook, 3rd ed., National Coal Board, 1983.

Velinsky, S.A., Anderson, G.L., and Costello, G.A. Wire rope with complex cross sections, *Journal of Engineering Mechanics, Trans. ASCE*, 110(3), 380–391, 1984.

Utting, W.S. and Jones, N. A survey of literature on the behaiour of wire ropes, *Wire Industry*, 51, 623–629, 1984.

Jones, N. Behaviour of wire ropes in offshore applications, *Wire Industry*, 51, 471–473, 1984.

Raoof, M. and Hobbs, R.E. The bending of spiral strand and armored cables close to terminations, *Proceedings of 3rd Offshore Mechanics and Arctic Engineering Symposium, ASME*, New York, 1984, Vol. 2, pp. 198–205. Also to appear in ASME, Jnl of Energy Resources Technology, 1984.

Velinsky, S.A. General nonlinear theory for complex wire rope, *International Journal of Mechanical Sciences*, 27 (718), 497–507, 1985.

Velinsky, S.A. Analysis of fiber core wire rope, *Journal of Energy Resources Technology, Trans. ASME*, 107(3), 388–393, 1985.

Phillips, J.W. and Costello, G.A. Analysis of wire ropes with internal-wire-rope cores, *Journal of Applied Mechanics, Trans. ASME*, 107, 510–516, 1985.

Utting, W.S. and Jones, N. Tensile testing of a wire rope strand, *Journal of Strain Analysis*, 20(3), 151–164, 1985.

Lanteigne, J. Theoretical estimation of the response of helically armored cables to tension, torsion and bending, *Journal of Applied Mechanics, Trans. ASME*, 52, 423–432, 1985.

Utting, W.S. and Jones, N. The response of wire rope strands to axial tensile loads—Part I. Experimental results and theoretical predictions, The response of wire rope strands to axial tensile loads—Part II. Comparison of experimental results and theoretical predictions, *International Journal of Mechanical Sciences*, 29(9), 605–636, 1987.

Chien, C-H. LeClair, R.A., and Costello, G.A. Strength and fatigue life of wire rope, *Mechanics of Structures and Machines*, 16(2), 213–223, 1988.

Velinsky, S.A. and Schmidt, J.D. A simplified treatise on the effect of wear in cables, *Journal of Offshore Mechanics and Arctic Engineering, Trans. ASME*, 110(1), 32–37, 1988.

Velinsky, S.A. Design and mechanics of multi-lay wire strands, *Journal of Mechanisms, Transmissions and Automation in Design, Trans. ASME*, 110(2), 152–160, 1988.

Raoof, M. and Hobbs, R.E. Analysis of multilayered structural strands, *Journal of Engineering Mechanics, Trans. ASCE*, 114(7), 1166–1182, 1988.

LeClair, R.A. and Costello, G.A. Axial bending and torsional loading of a strand with friction, *Journal of Offshore Mechanics and Arctic Engineering, Trans. ASME*, 110, 38–42, 1988.

Cappa, P. An experimental study of wire strains in an undamaged and damaged steel strand subjected to tensile load, *Experimental Mechanics*, 28(4), 346–349, 1988.

Index